Research Methodology

Research Methodology

A Practical and Scientific Approach

Edited by Vinayak Bairagi
and Mousami V. Munot

CRC Press
Taylor & Francis Group
Boca Raton London New York

CRC Press is an imprint of the
Taylor & Francis Group, an **informa** business

A CHAPMAN & HALL BOOK

CRC Press
Taylor & Francis Group
52 Vanderbilt Avenue
New York, NY 10017

© 2019 by Taylor & Francis Group, LLC

CRC Press is an imprint of Taylor & Francis Group, an Informa business

No claim to original U.S. Government works

Printed on acid-free paper

International Standard Book Number-13: 978-0-8153-8561-5 (Hardback)

This book contains information obtained from authentic and highly regarded sources. Reasonable efforts have been made to publish reliable data and information, but the author and publisher cannot assume responsibility for the validity of all materials or the consequences of their use. The authors and publishers have attempted to trace the copyright holders of all material reproduced in this publication and apologize to copyright holders if permission to publish in this form has not been obtained. If any copyright material has not been acknowledged please write and let us know so we may rectify in any future reprint.

Except as permitted under U.S. Copyright Law, no part of this book may be reprinted, reproduced, transmitted, or utilized in any form by any electronic, mechanical, or other means, now known or hereafter invented, including photocopying, microfilming, and recording, or in any information storage or retrieval system, without written permission from the publishers.

For permission to photocopy or use material electronically from this work, please access www.copyright.com (http://www.copyright.com/) or contact the Copyright Clearance Center, Inc. (CCC), 222 Rosewood Drive, Danvers, MA 01923, 978-750-8400. CCC is a not-for-profit organization that provides licenses and registration for a variety of users. For organizations that have been granted a photocopy license by the CCC, a separate system of payment has been arranged.

Trademark Notice: Product or corporate names may be trademarks or registered trademarks, and are used only for identification and explanation without intent to infringe.

Library of Congress Cataloging-in-Publication Data

Names: Bairagi, Vinayak, editor. | Munot, Mousami V., editor.
Title: Research methodology : a practical and scientific approach / editors, Vinayak Bairagi, Mousami V. Munot.
Description: Boca Raton : CRC Press, Taylor & Francis Group, 2019. |
Includes bibliographical references and index.
Identifiers: LCCN 2018046891| ISBN 9780815385615 (hardback : alk. paper) | ISBN 9781351013277 (ebook)
Subjects: LCSH: Research–Methodology.
Classification: LCC Q180.55.M4 R46445 2019 | DDC 001.4/2–dc23
LC record available at https://lccn.loc.gov/2018046891

Visit the Taylor & Francis Web site at
http://www.taylorandfrancis.com

and the CRC Press Web site at
http://www.crcpress.com

Dedicated to

the contributing authors and our well

wishers

Contents

Preface ix
Contributors xi

1. Introduction to Research 1
   Geetanjali V. Kale and J. Jayanth

2. Literature Survey and Problem Statement 25
   Dr. Manoj S. Nagmode

3. Research Design 69
   Prachi Joshi

4. Basic Instrumentation 99
   Pradeep B. Mane and Shobha S. Nikam

5. Applied Statistics 161
   Varsha K. Harpale and Vinayak K. Bairagi

6. Presenting and Publishing the Research Findings 207
   Krishna Warhade, Vinayak K. Bairagi, and J. Jayanth

7. Plagiarism 237
   Mousami V. Munot, Sesha S. Srinivasan, and Anand S. Bhosle

8. Intellectual Property Rights 265
   Dipali Kasat, Rajeev Kumar, and Shailaja Patil

Index 291

Preface

"Research" is no longer confined just to students and academics but is an important component for all professionals and promising experts (both offline and online) in every domain. Any investigation involving validation of facts, revealing intriguing results, and/or new actualities demands exhaustive research to ensure excellence. Research methodology includes a wide assortment of tools and techniques for systematic and effective research. This book, *Research Methodology: A Practical and Scientific Approach*, presents the essential fundamentals and directional pointers to researchers and postgraduates to enable them to initiate research, lay foundations, proceed methodically, and continue steadily. This recommended textbook contains precise reading material for both researchers and their research guides/supervisors. The editors of the book and the authors of each chapter have unanimously made a sincere attempt to provide a blend of technical and nontechnical language, test cases, and samples, in order to ensure easy reading and understanding. Starting in the first chapter with a brief introduction to research and research methodology, the book highlights other dimensions of research including surveys, problem formulations, research design, the importance of statistics, instrumentation, plagiarism, and intellectual property rights through numerous practical examples, tools, and Web sources. Various scientific methods and basic postulates are also detailed. This book will prove useful to all researchers irrespective of field. This book is also beneficial to students pursuing their postgraduate degrees and examinations.

Dr. Vinayak K. Bairagi
Dr. Mousami V. Munot

Contributors

Vinayak K. Bairagi completed the M.E. (Electronic) at Sinhgad COE, Pune, in 2007 (1st rank in Pune University). The University of Pune awarded him a Ph.D. degree in engineering. He has teaching experience of 12 years and research experience of 8 years. He has filed 10 patents and 5 copyrights in the technical field. He has published more than 58 papers, of which 26 are in international journals. He is a reviewer for nine scientific journals, including *IEEE Transactions, The IET Journal,* and other Springer journals. He is the P.I. for a UoP-BUCD research grant. He has received the "Maniratna" Best Teacher Award for Excellent Academic Performance (2013). He has received an IEI national level Young Engineer Award (2014) and an ISTE national level Young Researcher Award (2015) for his excellence in the field of engineering. He is a member of INENG (UK), IETE (India), ISTE (India), and BMS (India). He worked on image compression at the College of Engineering, Pune, under Pune University. Currently he is associated with the AISSMS Institute of Information Technology (affiliated college to S. P. Pune University), Pune as a professor in electronics and telecommunication engineering. He is a recognized Ph.D. Guide in electronics engineering at Savitribai Phule Pune University. Presently he is guiding seven Ph.D. students.

Anand S. Bhosle is currently working as an assistant professor in the Department of Electronic and Telecommunication at the Pune Institute of Computer Technology, PICT, Pune, India. He completed the B.E. in electronics and communication engineering from VTU (Visvesvaraya Technological University), Belgaum Karnataka. He obtained his M.Tech. in communication systems from Rashtreeya Vidyalaya College of Engineering (RVCE), Bangalore. He has over six years of teaching experience and two years of industrial experience. His research interests include wireless networks and computer networks. He has published papers in five IEEE international conferences publications.

Varsha K. Harpale received the M.E. degree in electronics and telecommunication engineering (E&TC) from SPPU University, Pune, India, in 2009 specializing in in VLSI and embedded systems. She has teaching experience of 17 years and currently is pursuing her Ph.D. degree in E&TC engineering with a specialization in biomedical signal processing at the AISSMS Institute of Information Technology, SPPU University, Pune, India. She has published 19 papers and has received two best paper awards at IEEE and Springer conferences. She currently is working as an assistant professor in the department of E&TC at Pimpri Chinchwad College of Engg, SPPU, Pune, India. Her research interests include biomedical signal processing, processor design, and the development of real-time embedded systems.

Jayanth J. received a B.E. degree in Electronics and Communication Engineering from Vidya Vikas Institute of Engineering and Technology, Mysore, Karnataka, India, in 2008, an M.Tech. in Digital Electronics and Communication Systems from Malnad College of Engineering, Hassan, Karnataka, India, in 2010, and a Ph.D from VTU, Karnataka, India, in 2017. He has eight years of teaching experience. He has published 35 papers, of which 15 are in international journals. He has authored one book chapter and written three books on image fusion and image classification using fuzzy logic for remote-sensed data classification. He is a Co-PI for the DST NRDMS project research grant. He has completed three consultancy projects for Infy-cloud Bangalore. His subjects of interest include image processing, swarm intelligence, and remote sensing. He is a member of ISTE. Currently, he is working with the GSSS Institute of Engineering & Technology for Women, Mysuru, Karnataka (affiliated with the Visveswaraya Technological University (VTU), Belagavi, Karnataka), as an associate professor in electronics and communication engineering. He is a recognized Ph.D. Guide in electronics engineering.

Prachi Joshi completed her Ph.D., M.Tech, and B.E. in computer engineering at College of Engineering, Pune (COEP), the University of Pune. She has more than 10 years of teaching and research experience. She is co-author of *Artificial Intelligence*, published by PHI; she also has contributed chapters to a book on big data analytics. She has successfully supervised many projects at the graduate and post-graduate level in the areas of artificial intelligence, data mining, and machine learning. Her research interests include information retrieval and incremental machine learning. She also maintains a blog on machine learning. She worked at the MIT College of Engineering as an associate professor in the Department of Computer Engineering. She is currently a freelancer and is actively involved in conducting workshops on data analytics.

Geetanjali V. Kale completed her B.E. (Computer Science and Engineering) with Dr. Bamu in 2001 and her M.E. (Computer Engineering) at the Pune Institute of Computer Technology in 2004. She completed her Ph.D. in Computer Engineering in 2017 from Savitribai Phule Pune University. She has 16 years of teaching experience. She has filed one patent and published more than 18 papers in international journals and conference proceedings. She also has worked as a reviewer for several conferences and journal papers. She was the P.I. for a UoP-BUCD research grant. She is member of ACM and a lifetime member of ISTE. Her area of interest include computer vision, machine learning, and data science. She is associated with the Pune Institute of Computer Technology (affiliated with Savitribai Phule Pune University) as an assistant professor in computer engineering.

Dipali Kasat is an M.Tech. from VIT, Pune, Maharastra. She completed her Ph.D. from Amravati University, Maharashtra. She has teaching experience of 15 years and research experience of 8 years. She has filed two patents and two copyrights registered in the technical field. She has published more than 15 papers. She co-authored *Management Information System*. She is a reviewer for scientific journals, international conferences, and Ph.D. theses at the universities of Amravati (Maharashtra) and Jhunjhunu (Rajasthan). She is the resource person for IPR at the Gujarat Council for Science and Technology(GUJCOST). She was director of Gujarat Technology University's Innovation Sankul project for the Surat region. She is the member of IETE (India), ISTE (India), and CSI(India).Currently she is at the Sarvajanik College of Engineering and Technology (affiliated college of Gujarat Technological University), Surat, Gujarat, as an associate professor in the computer engineering department, and she heads the IP cell and contributes to IPR culture promotion among students.

Rajeev Kumar is an assistant professor in the Department of Environment Studies, Panjab University, Chandigarh. He completed his B.Sc. and M.Sc. (Honours School) degrees in chemistry from the Centre of Advanced Studies in Chemistry, Panjab University, Chandigarh. He also received his Ph.D. in organomettalic chemistry with

Prof. K. K. Bhasin and Prof. S. K. Mehta there. He has published seven book chapters and 21 research publications in international journals. His publications have more than 150 citations. He currently is working with six Ph.D. students; one student completed her Ph.D. in March 2018. He received the MASHAV Fellowship from the Ministry of Foreign Affairs, Israel, in 2017.

Pradeep B. Mane received his B.E. (E&TC) and M.E. (E&TC) from the Government College of Engineering, Pune, India, and his Ph.D. from Bharati Vidyapeeth University, Pune. He worked for Philips India Ltd., Bharati Vidyapeeth COE Pune, and currently is working as a principal at AISSMS's Institute of Information Technology, Pune, affiliated with Savitribai Phule Pune University. He was a member of the BOS for the electronics faculty at Bharati Vidyapith University and Savitribai Phule Pune University. He has co-authored six books for engineering courses on radio and TV engineering and computer networks. He has published 50 papers and attended national, international conferences, and seminars. He has 47 publications in international journals. He is a regular reviewer for *Wireless Personal Communication*. His area of interests are wired and wireless communication and computer networks. He was CO-PI for an ISRO-UOP research grant. He is a fellow of IEI and IETE and a member of IEEE, ISA, IJERIA, and ISTE. He has received national awards for the Best Engineering College Principal (2017) from Bharatiya Vidya Bhavan (ISTE) and the Computer Society of India (CSI). He is recognized as a Ph.D. Guide for Savitribai Phule Pune University and Bharati Vidyapeeth deemed university in Electronics engineering. Three students have completed their Ph.D. under his guidance and eight students are currently working with him.

Mousami V. Munot is currently an associate professor in electronics and telecommunication department at Pune Institute of Computer Technology, PICT, Pune, India. She has 13 years of teaching experience and 7 years of research experience. She completed her M.Tech. (1st rank holder) from the College of Engineering, Pune, CoEP, in 2007. Savitribai Phule Pune University, SPPU, awarded her a Ph.D.in 2013. She received a fellowship from the Indian National Academy of Engineering (INAE) for research work under Dr. Jayant Mukhopadhyay, Department of Computer Sciences, IIT, Kharagpur, India. She also received support from DST (ITS) to present a research paper at IET – Image Processing Conference, London, UK. She has received best paper awards at national and international conferences. She has received grants worth 20 Lakhs from DST, AICTE, and SPPU-BCUD for various projects. She has served as session chair, IEEE-WIECON, AISSMS, Pune, India, 2016; a TPC member, ICDECT, 2016; Lavasa City, Pune-India, 2016; session chair, EEECoS, 2016; AISSMS, Pune, India, 2016; and as guest editor for a special issue on soft computing in the *Journal of Electrical and Electronics Engineering* (January 2015). She is a reviewer for several journals, including *IEEE, System, Man and Cybernetics, Springer Medical and Biological Engineering Computing, Springer Journal of Medical Sciences, Canadian Journal of Electrical and Computer Engineering*, and *IET –Image Processing*. She is a member of IET, ISTE, BMESI, and IoE..

Manoj Nagmode has a Ph.D. (Electronics and Telecommunication Engg) from the Government College of Engineering, Pune (2009) from Savitribai Phule Pune University (formerly known as University of Pune). He has teaching experience of more than 21 years. He has published more than 40 papers, 26 of which are in international journals, six of which are in SCI indexed journals. He has worked as an expert for various research review committees. He has published four books. He is a reviewer for several scientific journals. His research domain is in the area of signal and image processing, video processing, VLSI design, and embedded systems. He is a member of ISTE (India) and ISOI (India). Currently he is associated with the Government College of Engineering and Research, Avasari (established by the Higher and Technical Education Department, Government of Maharashtra, and affiliated with S. P. Pune University), as a Professor and Head of the Department in Electronics and Telecommunication Engineering. He is a recognized Ph.D. guide in Electronics Engineering at Savitribai Phule Pune University. Presently he is guiding six Ph.D. students.

Shobha S. Nikam received her B.E. (E&TC) degree from SVPMS's College of Engineering, Malegaon, India, in 2007 and her M.E. (E&TC – Microwaves) from AISSMS's College of Engineering, Pune, in 2012. She is currently pursuing her Ph.D. at AISSMS's Institute of Information Technology, Savitribai Phule Pune University, Pune. She has teaching experience of 10 years and research experience of 1.5 years. Currently she is associated with the AISSMS Institute of Information Technology (affiliated college to Savitribai Phule Pune University), Pune, as an assistant professor in electronics and telecommunication engineering. She has published papers in 13 international journals and presented papers at five international conferences. She was the principal investigator for an ISRO-UOP research grant. She is a member of IEEE, ISTE, and IEIE.

Shailaja Patil is currently working as a professor in the Department of Electronics and Telecommunication, and Dean (R&D) at the Rajarshi Shahu College of Engineering, Pune, India. She has 21 years of teaching and 3 years of research experience. She received a D.Phil. in computer engineering from Saradar Vallabhbhai Patel Natioal Institute of Technology, Surat, in 2013. She has carried out research on statistical techniques in localizationa and tracking using wireless sensor networks. She completed her P.G. Diploma in Patent's Law at Nalsar Law University, Hyderabad, in 2018. She has authored *Feedback Control System*. She has six patents and four copyrights to her credit and more than 60 publications in national and international journals and conferences proceedings. She is a regular reviewer of *IET-Circuits, Devices & Systems*. She has served as session chair in several international conferences. She has received funding worth 8 Lacs from BCUD-SPPU and AICTE for Research and Faculty development programs. She is a Fellow of the Institution of Engineers and a member of various professional bodies, including IEEE, ISTE, GISFI, ISA, and ACM. Currently she is guiding seven Ph.D. students.

Sesha S. Srinivasan is currently an assistant professor of physics at Florida Polytechnic University (FPU), Florida, USA. Before moving to FPU in 2014, he was a Tenure Track Assistant Professor of Physics, at Tuskegee University (TU), Alabama, USA. Dr. Srinivasan has more than a decade of research experience in the interdisciplinary areas of solid state and condensed matter physics, inorganic chemistry, chemical and materials science engineering. His Ph.D. focused on the development various rare-earth, transition metals and intermetallic alloys, composites, nanoparticles, and complex hydrides for reversible hydrogen storage and environmental air/water purification applications. Dr. Srinivasan and his Ph.D. advisor, Professor O.N. Srivastava (BHU, Varanasi), have successfully converted a 4-stroke, 100 cc Honda motorcycle to run on hydrogen gas, which was delivered from the on-board metal hydride canister. He and his post-doctoral advisor from the University of Hawaii, USA, have extensively collaborated with scientists around the world for the hydrogen storage on lightweight complex hydrides, funded by the U.S. Department of Energy (DOE) and WE-NET, Japan. He has established a state-of-the-art research laboratory at the CERC and supervised several graduate and undergraduate students. He has also served as Associate Director of Florida Energy Systems Consortium (FESC) at USF to co-ordinate a number of research projects on clean energy and environment, which was funded by the State Energy Office Florida ($9M grant). Dr. Srinivasan has been awarded many research grants. He has recently awarded with two U.S. patents on hydrogen storage nano-materials' development and methodologies and one nonprovisional disclosure is filed on a recent technology in regard to waste water remediation. He has published eight book chapters and review articles, more than 85 journal publications, and many more peer-reviewed conference proceedings. Dr. Srinivasan has served as a reviewer in the panel review committee of the National Science Foundation (NSF), SMART and NDSEG panels of ASEE, ad-hoc merit review committee of US Department of Energy, and a panelist for Qatar National Research Fund (QNRF). Dr. Srinivasan currently serves as an associate editor of deGruyter Open Book publications in physics, materials science, and astronomy, and guest editor for a special issue of the *Journal of Nanomaterials, Hindawi Publications and Applied Sciences*. Dr. Srinivasan was a guest speaker on "Green Energy" at the 95th Indian Science Congress. He hosted a USA-India workshop-cum-conference on green chemistry/engineering and technologies graduate program at Florida Polytech.

Krishna Warhade completed B.E. (Electronics) and M.E. (Instrumentation) from Shri Guru Gobind Singhji Institute of Engineering and Technology Nanded, Maharashtra, India. He completed his Ph.D. at the Indian Institute of Technology Bombay (IIT Bombay), India. He has teaching experience of 22 years, including four years research of experience. He has published several papers in SCI journals and attended many conferences. He authored *Video Shot Boundary Detection*. Currently he is working as a professor in electronics and communication in the engineering department at MIT World Peace University, Pune, Maharashtra, India. His research interests include video retrieval, wavelets, digital signal processing, biomedical signal and image processing, filter design, and precision agriculture.

1
Introduction to Research

Geetanjali V. Kale and J. Jayanth

CONTENTS

1.1 Introduction 2
1.2 Objectives of Research 3
1.3 Motivation behind Research 3
1.4 Important Ingredients for Research 4
    1.4.1 Componental Theory of Individual Creativity 5
    1.4.2 Workgroup Creativity 6
1.5 Types of Research 6
    1.5.1 Basic Research 6
    1.5.2 Applied Research 7
    1.5.3 Descriptive Research 7
    1.5.4 Analytical Research 8
    1.5.5 Correlational Research 8
    1.5.6 Qualitative Research 8
    1.5.7 Quantitative Research 8
    1.5.8 Experimental Research 9
    1.5.9 Explanatory Research 9
    1.5.10 Exploratory Research 9
1.6 Phases of Research/Research Process 10
    1.6.1 Selection of Domain 12
    1.6.2 Formulating a Research Problem (Tentative) and
           Identification of Keywords 13
    1.6.3 Literature Review 14
    1.6.4 Redefining Research Problem, Objectives (Final) and
           Outcomes/Formulating Hypothesis 15
    1.6.5 Research Proposal 16
    1.6.6 Identifying Variable/Parameters and Research Design 17
    1.6.7 Data Collection and Representation 17
           1.6.7.1 Selection of Appropriate Method for Data Collection 18
           1.6.7.2 Selection of Appropriate Methods for Data Representation 19
    1.6.8 Testing of Proposed Design on Collected data/Hypothesis Testing 19
    1.6.9 Results and Analysis 20
    1.6.10 Research Report Writing 21
1.7 Research Methods versus Methodology 22
1.8 Features of a Good Research Study 23
1.9 Summary 24
Further Reading 24

> Research is what I'm doing when I don't know what I'm doing.
> —Wernher von Braun

> Research is to see what everybody else has seen and think what nobody has thought.
> —Albert Szent Gyorgyi

Learning objectives of this chapter include, to

- Understand basic terminology and fundamental concepts in research
- Comprehend basic flow of research process and formulate it for individual research
- Select and write problem definition in domain of choice
- Write research proposal for selected problem definition
- Analyze types of research methods and apply appropriate methods for defined problem
- Define research methodology for selected problems

This chapter will enable the researcher to:

- Initiate and systematically continue research work in a procedural way
- Define the research problem and propose the hypothesis
- Propose a research methodology for selected problems
- Select and apply appropriate research methods to research problems

1.1 Introduction

The word research is combination of "re" and "search," which means a systematic investigation to gain new knowledge from already existing facts. In other words, research is a scientific understanding of existing knowledge and deriving new knowledge that may be applied for the betterment of mankind. It is basically search for truth. Research contributes significantly to the progress of the nation as well as an individual with commercial, social, and educational advantages. Research is an important parameter to judge the development of any nation. Research is an important component of private and government sectors. Nowadays, interdisciplinary research is at high demand.

Research should always aim at providing efficient solutions to routine problems. Researchers should carefully choose the appropriate research method and follow a research process by referring to existing theories. Research differs from a traditional way of education, that is, learning concepts and writing the examination or performing activity. According to Clifford Woody, "Research comprises of defining and redefining problems, formulating the hypothesis for suggested solutions, collecting, organizing and evaluating data, making deductions and reaching conclusion and further testing the conclusion whether they fit into formulating the hypothesis."[1]

This chapter discusses the basics of research and will provide a bird's-eye view of the research process for novice researchers.

1.2 Objectives of Research

The main purpose of research is to find solutions to unsolved problems using scientific procedures and also to understand various phenomena scientifically. In addition, one of the major objectives of research is to find out a hidden, undiscovered truth. There are various objectives behind undertaking research by individuals as well as various organizations. However, there are some identified purposes for each research work. Some general objectives behind research include:

- Propose and test certain hypotheses that provide causal relationships between variables
- Discover and establish the existence of relationship, association, and independence between two or more aspects of a situation or phenomenon. (Such studies are known as correlational studies)
- Understand different phenomena and develop new perceptions about it
- Study and describe accurately the characteristics of situations, problems, phenomena, services, groups, or individuals. (This type of study/research is known as *descriptive research*)
- Explain unexplored horizons of knowledge
- Test reported findings and conclusions on new data and novel conclusions on previously reported data
- Study the frequency of research that is connected with unspecified study. (This type of study/research is known as *diagnostic research*)

1.3 Motivation behind Research

Research is a long process, so the main driving factor is motivation. For some researchers and post-graduate students, the main objective behind the research is to earn a degree. For organizations including defence and research laboratories, research is an important aspect. To philosophers and thinkers, research may mean the outlet for new ideas and insights, whereas to intellectual people research may mean the development of new styles and creative work. Irrespective of any domain, research demands passion.

Initially the research is a random walk (research scholar is not sure about topic/research problem), but one need to systematically continue to get the destination. Failure is an inevitable step in the research phase, for example, failure in getting results, publications, and so on. But the researcher's passion and motivation helps in such situations.

1.4 Important Ingredients for Research

Creativity, good written and verbal communication skills and in-depth knowledge of the subject are essential for successful completion of research work. Figure 1.1 shows important ingredients of research. A researcher should have sound fundamental knowledge of the domain to be undertaken. A querying attitude is one of the important factors. Anything and everything is questionable; this questioning attitude is essence of research and invention.

A creative person does things that have never been done before. Prof. Robert Sternberg has listed some "must have" characteristics in order to be creative. These skills are a combination of synthetic intelligence and practical intelligence. Practical intelligence is the ability to adopt day-to-day requirements. Persistent, tenacious, uncompromising, and stubborn are some of the characteristics of creative people. One should always challenge conventional solutions to a problem. Velcro and Post-it Notes, for example, are outcomes of creative thinking.

Some examples of creativity include:

- Apple iPhone, which convinced people for the value of replacing the keyboard with a touchpad, and touchscreen operations controlled by fingers as a replacement for a stylus. The invention of laptops, the Internet, supercomputers, and so on are some examples in the computer science and engineering domains
- Newton's law of motion and the universal law of gravity are some examples of creative inventions in physics
- Composing beautiful music and creative writing are examples of creativity in art
- Analyzing a situation in a different way is also creativity

FIGURE 1.1
Ingredients for a good researcher

- Analyzing history with a different angle and solving and analyzing cases in law with different perspectives are also creativity
- Putting philosophy differently is also creativity. For example, Charles Darwin's theory of biological evolution by natural selection is contradiction to the previous biological theories. However, Darwin's theory is base for many modern biological inventions

Creativity is a combination of generating ideas and validating it, including research, development, production, and marketing. Creativity is also considered to be a pre-innovation process; it is a prerequisite for scientific and technological innovation. Creativity is an inevitable component of any impressive action. Creativity is a quality that adds essence to the expected outcomes.

Important instances of creativity are to provide a novel work process through a valuable innovation. Scientific creativity is a source of innovation that is subdivided into four types, listed here:

- Set a new idea for formulation, which opens up a new cognitive frame to a new level of basic assumptions
- A new empirical formula should be discovered, which stimulates a new theory (for example, Darwin's theory of evolution)
- A new methodology should be developed to solve a theoretical problem that can be tested empirically
- A novel technique/instrument should be designed, which can propose a new technique with new possibilities

Creativity is observed at individual as well as group level. Detailing is given in subsequent section.

1.4.1 Componential Theory of Individual Creativity

The componential theory of individual creativity has three essential components of individual creativity:

- Expertise
- Creative thinking skills
- Intrinsic task motivation

Expertise comprises actually occurring knowledge, which consists of technical expertise with a special skill that can be rooted to distinct personality in the working group. Creative-thinking skill is "working with an unknown data with a greater extent than usual," found in people who work with an original idea to emphasize that there should be a separation between generating ideas and validating those ideas. Intrinsic motivation determines what a researcher should actually do and is influenced by environmental factors.

Creative process can be determined in a sequential way depending on preparation, incubation, illumination and verification. Incubation is a period during which a problem is banned from conscious thought, and solutions are generated in an unconscious way.

Triggering subconscious thinking during the incubation period is an effective approach of grooming creativity. Verification is a stage where the work is been tested for the validity of novelty and usefulness of the idea.

1.4.2 Workgroup Creativity

Research and development mainly generates scientific and technological information, which transforms the work into novel ideas and products through a process. Creativity in a group includes working in a laboratory with respect to organisational settings, which helps most scientific and artistic innovation through joint conversations as well as thinking that emphasizes the importance of the social dimension of creativity.

Knowledge is another very important ingredient. After getting an idea, one has to decide on the design and implementation. To improve on communication skills a researcher can take courses and listen to talks given by inventors and other researchers to understand how they communicate their ideas. Oral and written communication is integral part of research and it helps to convince people about your ideas, methodologies, and results.

Researchers should be focused on their own problem definition; however, they should always investigate whether the solution proposed fits in a global scenario. An advisor plays a very important role in research, specifically when pursuing a Ph.D.. An advisor guides the right path at different junctions and are torch bearers.

There is always scope for improvement in everything and one should always strive for excellence. Researchers should not neglect any of the skills required for research and continuously improve and strengthen those skills. In addition, researchers require qualities such as hard work, consistency, and patience. A researcher needs to be extremely committed to the research work to enjoy the fulfillment and rewarding moments.

1.5 Types of Research

Different research types are discussed in this section. They are classified in various categories including applicability, the mode of enquiry in conducting the study, and major objectives of the study. Selection of research method depends on the discipline of the research, objectives, and the expected outcomes. One research problem may use multiple research types. Main research types discussed in this section include basic research, applied research, descriptive research, analytical research, correlational research, qualitative research, and quantitative research (see Figure 1.2).

1.5.1 Basic Research

Basic research is pure or fundamental research; there is no immediate need, but new theories can be added to the knowledge cluster. This type of research may solve problems but may not have practical applications. It has a broader scope compared to applied research. Theories in basic sciences and mathematics are examples of basic research.

Introduction to Research

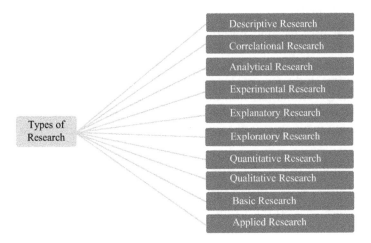

FIGURE 1.2
Types of research

Newton's laws of motion is an example of basic research. This has been applied in many product design and testing.

1.5.2 Applied Research

Applied research tries to solve an immediate specific problem faced by industry or society. The obtained solution can be deployed to solve the problem. The duration of applied research is shorter as a quick solution is expected. An optimized search problem on the Internet is an example of applied research in the computer engineering domain. "Analysis of cell/body organ behavior in cancer" is an example of hybrid research. A researcher may use data analytics, image processing, algorithms, and knowledge of the medical industry.

The outcome of applied research should either address the unsolved problem or improve the existing solution.

1.5.3 Descriptive Research

Descriptive research is generally used in business analysis or social problems. This type of research does not have any control over the parameters or variables. It just tries to represent or analyze the previous and or current facts.

Some of the examples of descriptive research include an analysis of customer purchase patterns, that is, purchases from the mall, online, or retailer, as well as the study of travel mode used by people. All kinds of correlational methods, survey methods, and comparative studies are descriptive research. Various kinds of systematic surveys are conducted, including the study of various cultural practices, the region-specific study of particular decease, and an analysis of the development of particular businesses. This is also known as ex-post facto research. Study of the effect of global warming on birds is also an example of descriptive research. One specific example is: "To study socioeconomic characteristics of residents of particular community during certain period for specific country."

1.5.4 Analytical Research

Analytical research uses existing information to explain a complex phenomenon or to perform a critical evaluation. The identified hypothesis can be accepted or rejected depending on the analysis; from experience the hypothesis can be redefined. Analytical research is observed in historical study, forensic work, food, in the medical domain, and so on.

Analytical research summarizes and evaluates the ideas in historical research for accessing both witness and literature sources to document past events. Philosophical research organizes data that can be presented to support the data in comprehensive model.

1.5.5 Correlational Research

Correlational research focuses on exploring the relationship or association between incidences, variables, and so on. Examples of correlational research include "To study the effect of a modern lifestyle on obesity" and "Analysis of the impact of technology on employment." In the first example, a modern lifestyle and obesity are two variables and researchers should study a group of people living both a modern lifestyle and a nonmodern lifestyle. The groups should be segregated on the basis of obesity parameters. Collected data may be analyzed to establish the relationship between two variables: "obesity" and "modern" lifestyle. Similarly, researchers need to study "employment" and "availability of technology." From the collected data, researchers may come up with number of observations and analytics.

An institute organizes a conference and takes feedback from participants as its regular practice, to provide input for evaluating the quality of current conference and to make improvement in next conference. Participants who attended the conference gave feedback stating that there was a difficulty in reaching the location. For the next conference, the survey question can include a question, "Was there a sufficient service available for reaching the location?"

1.5.6 Qualitative Research

Qualitative research mainly deals with the quality or the types of the parameters considered for the research. Here, it is assumed that the world is unstable and differences in the parameter may occur with time. Research related to human behaviour is an example of qualitative research. Everybody can react to the situation differently and it is difficult to propose the predictive conclusions. This type of research is more complicated and requires more guidance. Less emphasis is given on generalization and more focus is towards individual. An example can be "Study of behaviour of employees in an organization." Here, behavior of an employee may vary with different parameters such as gender, post, skill set, expertise, socioeconomic status, and religion. Focus of the work is to find results with respect to qualitative parameters. Another the example of qualitative research is, "How and why there is an upward movement in the value of dollar and its impact on the Indian currency."

1.5.7 Quantitative Research

Quantitative research involves measurements of quantities of characteristics that can be used as features for the research study. Unlike qualitative research quantitative research assumes that world is stable and uses statistical analysis on parameter values for conclusions. Statistical quantities that can be measured are involved in quantitative research. Example of quantitative research is "Finding number of individuals taking benefits of different government policies." This is a statistical report of various government policies and number of individuals and does not involve any qualitative parameter. Another example is to conduct a survey every weekend and a random question is been asked, "Who is your favorite actor?" and response of this questions is self-controlled by providing a multiple choice question. The outcomes are provided briefly in judgement form with statistical respective.

1.5.8 Experimental Research

Experimental research focuses on the fieldwork and experiments that can control the independent variable. Study of the effect of the new drug on a specific group of people or animal is an example of experimental research in medicine domain. In computer engineering, "Analysing performance of algorithms on various dataset" is an example of experimental result. A researcher wants to examine changes in the land cover using satellite data. In this example, to ensure the changes in the study area, researchers randomly place the area of changes including change in the agricultural land due to land degradation, NH roadways, and so on, which can be verified with quasi experimental research and can be tested and trained with pre- and postexperimental research design.

1.5.9 Explanatory Research

Explanatory research tries to analyze and justify the reason behind the occurrence of particular phenomenon or association between the variables. It basically answers the "Why" type of questions. It aims to explain why a relationship, association, or interdependence exist.

Explanatory research is also called as a causal research with a three important components like time-to-time sequences which will occur before the effect, concomitant variations, where the variations will be systematic between two variables.

Some of the examples of explanatory research are "Why the modernisation creates health problem?", "Why some students have casual attitude towards study, while others are sincere?", "Why will customers buy our food products in green packages?", "Which of the two advertising company will help students by providing the internship?", 'which company will be more effective and why?'

1.5.10 Exploratory Research

This kind of research generally explores the areas that have required meagre attention or it is for checking the possibility of research in the particular domain or area. A small-scale study is done to decide the further scope of advancement in domain. Depending on outcomes of exploratory study, domain is further explored for in depth research on the specific topic. Exploratory studies are also conducted to develop, refine and test procedures, policies, and tools. Some of the examples of exploratory research are, "Why product sales are reduced, due to already existing data or the products which have been acquired recently for the agricultural company?" Generally, the research statements that span across multiple domains come under exploratory research.

1.6 Phases of Research/Research Process

Efficient and well-planned activities always see success. Therefore, one need to efficiently plan a research activity, execute it meticulously, and publish it for outside world.

In a house-building task, the first blueprint needs to be finalized and one has to visualize all minute details such as position and size of rooms, doors, windows, balconies, and so on. Estimation of all required material should be finalized along with quality and quantity. Improper planning and estimates may not able to produce the expected dream house. Similarly, researchers should follow research process to get optimized research outcomes. General steps in the research process are shown in Figure 1.3 and selection of research problem is shown in Figure 1.4. Identification or selection of research domain is a very initial stage in the process of research, followed by formulating a tentative research problem. Selection of research guide can be done at initially or at any intermediate stages of research problem selection. If the problem is well-defined with research objectives initial phases can be combined and researcher can directly start with literature survey. Literature survey can be revisited number of times during research process. Researcher can switch from intermediate state to literature survey and back to the same state.

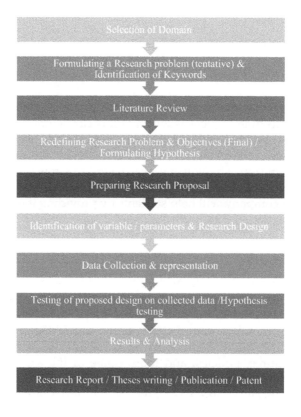

FIGURE 1.3
Phases of research

Introduction to Research

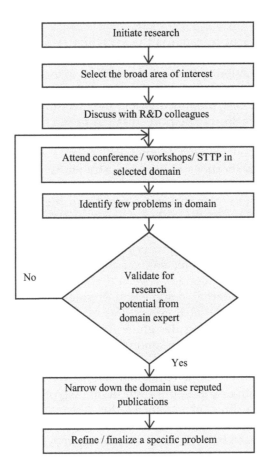

FIGURE 1.4
Phases of selection of research problem

General research process include:

- Selection of domain
- Formulating a research problem (rough) and identification of keywords
- Literature review
- Redefining research problem, objectives (final) and outcomes/formulating hypothesis
- Preparing research proposal
- Identifying variable/parameters
- Data collection and representation
- Testing of proposed design on collected data/hypothesis testing
- Writing and comparing results
- Research report writing

1.6.1 Selection of Domain

Selection of domain/area of research is the very first step in the process of research. There are different purposes behind undertaking particular research. If it is getting the degree, there is a high possibility to select the domain of the research guide or the guide may suggest selection of the particular domain. Otherwise, research domain can be selected from area of interest, identified gap in the literature and individual skill set. If the researcher is working in research organisation or laboratory the researcher has to work on the domain of organisation's interest. The purpose of this research may be to design and develop the new product, to upgrade available one, to study and analyse the effects of the product specifications in drug industry etc.

Every time, selection of brand new topic / domain for research is not required, in each field there are some topics that are old but still there is scope for research. In Electrical Engineering, domains like VLSI, Microelectronics, photonics, optics, digital signal processing, energy conversion has lot of material available still there is scope for more research. In Mechanical Engineering conceptualization, material analysis in design and manufacturing, thermal engineering, design of components (design of different engines like thermal engine), special tools and machines for manufacturing, electronically controlled internal combustion engine, diesel engine (reduce pollution), alternative fuel, taking real time pictures of machine and try to optimize the process. In Computer Engineering designing GPU / multi core enabled algorithms, management and analysis of large amount of data, data security and wireless communication in IoT devices are some of the areas of research. Interdisciplinary topics like IoT, application of genetic algorithms for optimization problems in different domain of engineering, application of analogy etc. are also open problems.

Even though there are different parameters for selection of domain, one should go with an area of his/her interest and current and future market trends. Researcher's interest is very important in basic research as outcome may not be immediately applied for product design or other social or commercial purposes. For applied research one should look at current and future requirements of market.

Austin O'Malley has well said that, "Well begun is not only half done but often fully cooked." Identification of an area of research and defining research problem are most important steps in the journey of research. One needs to give sufficient time and thought for the problem formulation. Before deciding domain or selecting problem one need to understand his/her passion, strengths, weaknesses, likes and dislikes. Area of research should be of your interest and it should have potential applications after completing your research. The motivation/reason behind the selection of research domain is very important. Some of the reasons for selecting particular research topic are the topic is of your interest, your research supervisor is expert in that domain and working on some related project, you have identified the gap in the literature, the topic has very good future prospects. In academic research, your supervisor may give you the topic or may ask you to select your own. While selecting your own research problem brainstorm the possible topics, verify the feasibility of a selected topic. Identify the important related keywords and review the literature with the identified keyword. Revisit the defined problem and rewrite if required. One should avoid selection of overdone topics unless you have some unique aspect for it. Also, a topic should not be too general and old. One should concentrate on current social or technical affairs of interest and keyword of that can also be used for research.

Introduction to Research

One invention gives rise to more opportunities for research in a particular domain. First Aeroplane designed by Wright brothers in 1903 was simple craft and if you look at today's aeroplanes, they are totally different. More sophisticated and efficient versions with modifications right from fuel consumption to the chair design for passenger's comfort. Over the period many engineers, researchers, and scientists from industry as well as academia worked on identified gap or problem faced during use and provided more efficient solutions than the existing ones. So, the problem can be totally new or it can be identified as a gap in the existing solution.

Researcher should choose a challenge suitable for his/her skill set. It should not be too high or too low in comparison with skill set.

1.6.2 Formulating a Research Problem (Tentative) and Identification of Keywords

For an in-depth literature survey researcher has to first define tentative research problem definition and identify the related keywords for literature search. Before starting with research; researcher should be equipped with all the tools and domain knowledge required for research. Research requires ability to go in depth of particular topic. The flowchart shown in Figure 1.4 provides general guideline for selection of research problem. Discussion of the state-of-the-art with colleague and domain expert is very important and helps a lot to finalize research problem. Attending workshops, conferences and or short term training programs is integral part of research work right from conception of problem to completion of work. Researcher should prepare the research tentative plan of research work. Table 1.1 shows the sample research plan

TABLE 1.1

Research Plan

Month (M) / Tasks	M1 to M6	M7 to M12	M13 to M18	M19 to M24	M25 to M30	M31 to M36	M37 to M42
Selection of Domain	■						
Formulating a Research problem (tentative) & Identification of Keywords	■						
Literature Survey	■	■					
Redefining Research Problem & Objectives (Final)/ Formulating	■						
Hypothesis	■						
Preparing Research Proposal	■						
Identification of variable/parameters & Research Design		■					
Data Collection & representation Analysis			■				
Testing of proposed design on collected data/Hypothesis testing				■	■		
Results & Analysis					■	■	
Research Report/Theses writing/Publication/Patent						■	■

prepared. Time span and tasks may change with domain, expert level of researcher, purpose of research, and so on.

Research statement should be clearly defined and its objectives should be SMART (Specific, Measurable, Achievable, Realistic, and Time-bound). Initially, one can start with tentative problem definition and review the state of the art literature related to topic. Subsequent section discuss in brief about literature review.

1.6.3 Literature Review

The literature survey is a comprehensive study of technical and authorised content related to research keywords. It is very important step in the initial phase of research, however, this step is revisited by researcher number of times during research journey. Literature survey provide details of research progress of particular domain. It helps the researcher to understand the approaches, methodologies, algorithms, and datasets used by other scientists. Also, it is important to identify where the gap is. Where can I contribute to an existing knowledgebase? It helps the researcher to understand the progress of domain and state of the art in the domain. Literature survey also helps to avoid duplication of work. One should always avoid reinventing the wheel; this is important if one of the purposes of the research is to get a PhD degree. Isaac Newton said, "If I have seen further than others, it is by standing upon the shoulders of giants." Researchers should able to use previously done work as foundation and propose new idea above that may be supporting or contradicting to previous work.

Open Access journals are one of the good sources of research articles, one can also access articles from library subscription, some university libraries etc. Good literature sources are periodicals, conference proceedings, research reports, standards, theses, dissertation reports, research reports, white papers, patents, reviews, textbooks, handbooks, encyclopaedia etc.

There are different strategies to search literature. One needs to first identify the keywords related to the problem definition and searching articles using keywords is the more precise way to search article. Backward and forward reference search is also popular and beneficial to understand the background of the research area. One can choose the latest article from the reputed journal written by good researcher and study all related references of that paper. This is a backward reference. One can also find the important old article in the domain and find all the papers that referred the article. This is forward reference search.

Novice researchers can follow below steps for literature survey-

(i) First download 100–120 research papers from published papers of IEEE/ACM/Wiley/Elsevier/Springer/T and F or some other good SCI Indexed Journals related to research topic and start reading it. Normally, we do this type of search with the help of keywords of research problems. Some of the search engines for the literature survey are Google Scholar, Citeseer, and so on.

(ii) Researchers can initially read only title, abstract and conclusion of each research paper and decide whether it is related to research topic. If yes, include in list, else skip it. From this step you will get to know "What has been done in this area and what can be done in future?" From this understanding one can come up with his/her research problem statement.

(iii) Read shortlisted papers in detail and start taking notes of different parameters considered, methodologies used, conclusions derived, and related mathematics/theories used. Researchers can finalize research hypothesis after this step and also can write good survey paper.

Another approach for literature survey is selecting any one best paper related to particular research problem of your interest. Try to find/download those papers that are cited and included in the references section. This approach leads to a better literature survey of a particular research problem.

During the literature survey process, researchers should able to identify the top 10 researchers in his/her domain. He/she should find good universities where people are working on the same or similar problems. Literature survey is a step that is often visited during the research journey. Researchers should always look at what other top researchers are doing in respective area of research.

Researchers should provide main focus on comparing the approaches, Identifying the weaknesses and strengths in recent research articles in the subject. Each paper's study can be summarized in systematic manner. Literature surveys can be done with respect to following points:

- What
- How (process schematic)/standard procedure (system block schematics)
- Discussion on major steps involved
- Design criteria and performance measures
- Techniques currently in use
- Comparative analysis (table or any suitable tool to discover and list pros and cons/ strength/weakness/future scope of existing techniques)
- Scope for research/gap in research

Researcher may add some more parameters to be studied about the research paper and can prepare a format for same to keep record of each paper.

Summarization and representation of literature surveys is an important part in the theses, presentations and in journal writing. Literature survey can be represented in textual, tabular or graphical format using different parameters.

1.6.4 Redefining Research Problem, Objectives (Final) and Outcomes/Formulating Hypothesis

Problem definitions should be unambiguous, clear statement that states the major objective of the research. There should be generally three to six subobjectives defined for research work. There should be a clear indication of the research work which should not be the recurrence of the same research. There should also be an outcome which has be initialized while mentioning the research objective. Objective should be given pointwise (three to six points).

It is important to define outcomes of research work to be carried out, which should show an impact of the proposed research work.

After domain selection and identifying the problem definition, researchers should formulate a hypothesis. State-of-the-art literature review on related topic creates a sufficient background for formulating the hypothesis. Reporting problem with the tentative solution is a hypothesis. The hypothesis is a tentative solution based on insufficient evidence that can be true or false. For example, "Algorithm A1 outperform over algorithm A2 for given data," "People who do not take balanced diet are obsessed than the people who take balanced diet." In the end, researchers either approve or disapprove the null hypothesis. Generally, it is easier to disapprove the null hypothesis as only one counter example is sufficient. On the contrary, exhaustive examination of all possible cases is required to prove the hypothesis.

If the outcome does not support the null hypothesis, we conclude with an alternate hypothesis. One can define a problem using null hypothesis as "There is no relation between the illness of children and change in season." If the result rejects the hypothesis, an alternate hypothesis is "Illness of children occurs mainly due to change in season." The null hypothesis is the precise statement about the parameters. Researchers either approve the hypothesis or disapprove the hypothesis. If researcher disproves the null hypothesis, all other possibilities are represented by alternative hypothesis. Null hypothesis does not provide a statistically significant relationship between variable, whereas, alternate hypothesis provides a statistically significant relationship between them.

1.6.5 Research Proposal

Researchers should able to convince people for selected topic and objectives through the research proposal. It is very important document that is reviewed by different committees. In industry proposal needs to be written for approval of top management, finance and marketing department. Research proposal is mandatory document to be submitted to the university or research organisation during the registration for a PhD degree. It can also be written to avail the funding from different agencies. Every agency generally has its own research proposal format. Research proposal formats generally include the following sections: Introduction (Proposed Topic of Research/Rational and Significance of the Study), Literature Survey (Background of the Proposed Research/Study of Research Work Done in the Area and Need for More Research), Motivation, Research Statement, Objectives of the Research Proposed, Probable Methodologies/Techniques to be Used, Expected Outcome(s) (the kinds of conclusions expected and their possible value), Plan of Research Work, References/Bibliography.

- **Introduction (250–300 words)/Proposed Topic of Research**
  In this part, the introduction to the area of interest can be included along with its potential for research and its application. Introduction also specifies the work to be proposed and it should be related to the title of the theses in case of the PhD dissertation.
- **Literature Review of Research Topic (1,500–2,000 words)**
  Present status of topic of research from existing literature can be written in this part of proposal. Literature of our work is been prepared by the scholar after going through the contemporary literature done in the area of interest after referring the Science Citation Indexed (SCI) Journals/e-journal, reputed conferences, magazines, MTech and PhD dissertations, reports published in the institution and company organisations along with patents in domain of research. In this context there should be a broad summary that should reveal the present status in the area of work and should highlight the identified research gap.

- **Gap in Existing Research**
  Issues that are unresolved on the research topic should be reported in the project with reference to the current status. Furthermore, possible utilization of the research outcome can also be suggested.
- **Objective of the Proposed Research**
  It should be clearly indicating the perception of the research work and should not be a mere repetition of the topic of research. What is to be achieved as an outcome of the research has to be visualized while mentioning the objective of the research. Objectives should be given pointwise (three to six points).
- **Outcome**
  It is important to make clear the impact of the proposed research and the particular aspect of the problem that is anticipated to produce an original contribution(s) by the candidate.
- **Methodology**
  Methodology should provide the experimental/infrastructural/computing facilities needed to carry out the research work and it should be described in the logical phases that are used to investigate the identified research problem. There should be a phasewise description with a brief explanation under each phase. The methods and approach can also be represented by figures and flowcharts.

1.6.6 Identifying Variable/Parameters and Research Design

The variable is basic quality or attribute that differs in value under different circumstances. The researcher should identify all related variables or parameters. Parameters can be identified during literature survey and it may vary depending on the proposed hypothesis. In hypothesis "Illness of children occurs mainly due to change in season," parameters can be age, whether ill, duration of illness, type of illness, season during illness, weight, height, temperature, and so on. These are some of the variables for considered hypothesis. One need to define the domain and range of each variable. Its range can vary with application. Variable can be either dependent or independent. Independent variables show changes in dependent variables. It may possess either continuous or discrete values.

To prove or disprove the hypothesis one needs to decide on the design strategy of research. The researcher needs to shift the paradigm from "what is my research?" to "how am I going to conduct it?" The research design is a systematic plan designed to obtain a solution to the research problem. Which data should be tested, which procedure/algorithm/methodology will be applied is decided in the design phase.

Research design should be carefully planned and checked for feasibility. It is blue print of your research. Good research design provides maximum outcome with minimum efforts. The researcher should think about size and type of samples and dataset to be considered, parameters to be considered. Researcher can revisit the literature review step again to check with state of the art design methodologies used by other researchers and can come up with methodology suitable for his/her own research hypothesis.

1.6.7 Data Collection and Representation

Data can be either directly collected afresh known as primary data or already collected and used data is known as secondary data. Secondary data is already considered by researchers to test the hypothesis. Most of the times in experimental research data is collected afresh to test the methodologies, hypothesis or algorithms. If researcher decides to use afresh data, then he has to decide carefully on method of data collection. Depending on problem definition and research objectives some parameters needs to be controlled during the data collection. In image or video processing applications related research topics researcher may control the parameters like frame rate, image resolution, illumination conditions, colour, number of cameras, and so on. These different parameters can be either kept constant or included in the study.

Consider the research topic "Study of effect of exercise on thyroid patient." Persons having hypothyroid or hyperthyroid are of interest and normal individuals are not. Thyroid level, duration of exercise, type of exercise, age, eating habits, and so on are different parameters that need to be taken into consideration during research.

Data can be represented in simple text, tables, graphs, audio, video, or images. Researchers should consider sufficient amount of data with all parameters. An incomplete dataset may not give a conclusive result.

Collected raw data should be first examined for errors, this is called editing of data. It should be coded and represented in user friendly format.

1.6.7.1 Selection of Appropriate Method for Data Collection

Nature, Scope and objective of enquiry constitute the most important factors affecting the choice of a particular method.

Funding availability decides the method for collection of data and the disposal of the research in a limited version. So selecting a cheaper method may not be efficient and effective but as to act within its limitation.

Time availability is also needed to be taken into account for deciding the method of collecting the data and takes more time when compared with shorter duration methods. Thus affects the selection of the method for processing the data. Another important thing that should be considered is the time precision for collection of the data.

The reliability can be tested by finding out the following:

- Who collected the data?
- What were the sources of the data?
- Were they collected by using proper methods?
- At what time were they selected?
- What level of accuracy was desired?
- Was accuracy achieved?

The data that are suitable for one enquiry may not necessarily be found suitable in another enquiry. In this context, the researchers must carefully scrutinise the definition of various terms and units of collection used at the time of collecting the data from the primary source originally. If the researchers find the difference in these, the data will remain unsuitable for the present enquiry and should not be used.

If the level of accuracy achieved in data is found inadequate for the purpose of the present enquiry, they will be considered as inadequate and should not be used by the researcher.

1.6.7.2 Selection of Appropriate Methods for Data Representation

Tabulation is a process that conserves space and reduces explanatory and descriptive statement and also provides the comparison from one state to another. Tabulation also provides various statistical computational information, which facilitates the summation of items and detection of errors and omission.

Importance of Tabulation

- Each table will give a clear title that does not require an explanation
- Each table will be provided with a separate number that will be easy for referring

Tables can be categorized in a chronological, geographical, and alphabetical order. Table should be made as logical, clear, accurate, and simple as possible. Most of the research outcomes consist of data that are of same kind; these types of data are needed to be reduced for the better understanding of the result. Classification of the data depends on the nature of a fact that is involved. Classification can be categorised under two sections such as numerical and descriptive. Descriptive refers to qualitative phenomenon which cannot be measured quantitatively and depends on the specified attributes. Whenever the data is classified according to the attributes, researchers should give a clear picture related to the attributes.

The numerical classification refers to the phenomenon which will measure the data quantitatively using statistics. Data related to the salary, classes of age, and so on, comes under these classes. Such data are known as statistics of variables and are classified on the basis of class interval.

Graphical representation helps to understand the data easily. All statistical package, offers a wide range of graphs. In case of qualitative data, the most common graph is bar chart and pie chart.

1.6.8 Testing of Proposed Design on Collected data/Hypothesis Testing

Researcher defines the hypothesis and he/she needs to test that hypothesis to prove or disprove. Hypothesis defining is discussed in Section 1.6.4, this section focus on hypothesis testing. Hypothesis testing is expressed as either a null hypothesis or alternative hypothesis. If the researcher compares two methods P and Q for superiority and if proceed on the assumption that both methods are equally good, then this assumption is termed as the null hypothesis (H_0). If the researcher thinks that "Method A is superior and the method B is inferior," then it is termed as alternative hypothesis (H_a). There are Type I and Type II errors related to the null hypothesis. We may reject H_0 when H_0 is true. This is a Type I error. Type I error means rejection of hypothesis that should have been accepted. Type II error means accepting the hypothesis that should have been rejected. The researcher may accept H_0 when H_0 is not true. α and β denotes the probability of Type I error and Type II error, respectively. Curve with conditional probability of rejecting H_0 as a function of population parameter and size of the sample is known as power curve of hypothesis testing. In short it is a plot of values of (1-β) for each possible value of population parameter for which the H_0 is not true. This curve is defined by a function known as power function. Some of the important limitations of the discussed test are (i) the result cannot be expressed with full certainty, they are probabilistic, and (ii) testing is not a decision-making activity in itself, but the researcher should not use it in a mechanical way.

Validity of the hypothesis should be tested on the basis of the data collected by the researcher. The procedure for hypothesis testing refers to all of those steps that we undertake for making a choice between the rejection and acceptance of a null hypothesis. The various steps involved in hypothesis testing are as follows:

- Setting up of hypothesis consists of the data that makes the statement of a null hypothesis, which should clearly state the nature of the research problem
- Particular expression of the hypothesis is an important aspect while considering a goal or purpose of the considered problem
- Hypothesis can be validated when the values are decided in advance for the significance of the work when they are directional and nondirectional
- Test statistics will be conducting hypothesis test for means and variance. The formula for test statistics and their distributions are discussed depending on the value of test statistics using observations selected by the researcher and the parametrical value stated under null hypothesis
- Using different type of critical value for test statistic, level of significance, and the type of test we obtain a critical value
- The null hypothesis is rejected or accepted by comparing the distribution of test statistics

Some of the important limitations of the discussed test are

- Results cannot be expressed with full certainty; they are probabilistic
- Testing is not a decision-making activity in itself, the researcher should not use it in a mechanical way
- Tests don't explain the reason why the dissimilar result has been obtained due to fluctuation
- Significance of the results is been validated on the basis of the probabilistic conditions which cannot be explained fully
- Inference the statistical data cannot provide the evidence for the truth of the hypothesis

There exist the number of statistical tools like t-test, F-test, chi-square test, D-W test, etc. to test the validity of the hypothesis.

1.6.9 Results and Analysis

The result is an important section of the research. If the topic is totally new then one cannot compare his/her results with existing. Here, the hypothesis should be tested with multiple approaches and that results can be compared. If already some scientists worked on methodologies or the topic it is important to compare your results with the existing state of the art results from the literature. While comparing results one should look at the considered parameters, dataset, and so on. Results should be represented in the visual format using tables, figures and or graphs. It should be properly labelled, clear and easy to understand. Carefully choose the type of graph taken for result representation. The relationship between the data in tables/graphs should be explained along with observations. Result section may include the problems faced during collection of data and complete analysis of results.

1.6.10 Research Report Writing

There are different purposes for writing the research report. The research report is a medium to convey research outcomes, contributions, findings and results to the outside world. It decides the quality of your research work. Without approval of experts in domain, the research is incomplete. Publication of research in open, referred international journal is very important aspect of the research. It can be research papers written for publication through different agencies like IEEE, Springer, Elsevier, T&F, and so on. Research report can be in the form of the theses for academic research and it is compulsory and important part of the completion of the degree. Patents, copyrights, and white papers are also possible outcomes of the research. Communicating ideas is important aspect of research. People should agree that your work is creative and novel. For good writing, lot of good reading is recommended to understand how people communicate their ideas.

Different agencies provide their own formats for writing the report. Format and length of the report depend on the type agency.

The researcher should first start with an introduction and can continue with literature survey, methodology, datasets, and results. Then conclusion section can be added and at last abstract of the paper, report or theses can be written. These are general guidelines suggested for novice researchers and not hard and fast rule. These may include the following sections:

- Abstract
- Introduction
- Review of Literature
- Problem Definition and Objectives
- Methodology
- Observations and Results
- Discussion
- Summary
- Conclusion
- Publications
- Bibliography/References
- Appendixes

This is a general guideline and every university or organization may have their own format of theses.

Researchers should be aware of plagiarism issues, copyright issues, procedures, and penalties. Plagiarism is to literally steal the work of another person, presenting it as original research without proper citation. There are different agencies including Turnitin, Unkund that verify the copied content in written document with its source. After completion of writing, a plagiarism check of the written document by an authorized agency is recommended.

While reporting own work, if the researcher is referring to work by other scientists or researchers, all of this should be acknowledged and added to the references. References can be organized in alphabetical or chronological order. Acknowledgments to people, funding agencies, or other organizations that helped directly or indirectly in the completion of the research work should be added.

The writer should have good command over the language especially the grammar of the language in which the work is reported. In science and technology domain research work is generally reported in the English language.

At glance during research process researcher should monitor following stage-wise details,

- **First stage**
    - Where do you start?
    - Area of interest should be decided
    - Check out why this area is of interest
    - Discuss your idea with local R&D staff

- **Second stage**
    - From the existing source systematic reviews should be considered carefully before starting the research
    - Duplication of research which is not of sufficient quality is itself unethical. So working/reviewing with existing knowledge is an important step

- **Third Stage**
    - Research domain should be determined
    - Data should be arranged for a specific purpose and statistically analysed
    - Determine top-down or bottom-up approach or combination
    - Data should be compared and validate on the work done
    - Data should provide the information and action performed for comparison of the work
    - Results should dealt with a means of solving the problem
    - Research validation should be verified with surveys
    - Perform research validation using research methods (interviews, surveys, etc.)
    - Appropriateness of the work should be justified
    - Cost should be specified
    - Test Practicality

1.7 Research Methods versus Methodology

Researcher should understand the difference between research method and research methodology. Research methodology explains more about the research process whereas research methods aim at finding answers to research questions. All of the methods, approaches, processes and techniques used by researcher during research process can be referred as research methods. It can be methods applied during dataset creation, data preprocessing, and data collection to decide sufficiency of data. Researcher has to use different analytical tools and techniques at the time of statistical analysis and to check accuracy of obtained results. Examples of methods applied during dataset creation are survey methods, interview method, questionnaires, thorough observations, and input through mechanical/electrical devices. Some of the examples of statistical analysis methods are mean, mode, median, mean deviation, standard deviation, skewness, kurtosis, and histogram. Methods such as finding precision, recall, f-score, and t-test are used in result analysis.

Researchers should be able to identify appropriate relevant methods applicable at each step. For researchers, along with an understanding of different research methods, design of methodology for his/her research problem is very important and it may differ from problem to problem. Consider the example of building a house; first, the architect will design a plan for your house according to your requirements. In his/her plan he fulfils each and every requirement and comes up with final solutions including positions of different rooms, windows, doors, staircases, and so on. The quality and quantity of each and every material is decided before constructing actual house. Estimation of building house is prepared before starting actual work. Then contractor decides how the construction should be completed, which tools and methods should be used at every stage of progress and plan is executed accordingly. First, complete plan of process of construction is prepared and then executed using tools and techniques. Similarly, in research researcher is expected to prepare plan of research from problem definition, that is, methodology for research process. At the time of deciding on a specific methodology, researchers should look at different available methods, analyse them and should choose the appropriate one for his/her research. In short, different research methods are part of research methodology.

1.8 Features of a Good Research Study

Related state-of-the-art literature should be studied in depth to avoid reinvention of wheel. Good research should clearly define the methodologies used, which should be replicable. It should be time-bound and realistic. Good research should have systematically chosen methodologies and datasets to prove the proposed hypothesis. Validity and reliability of data should be checked and researchers should consider an adequate amount of data. Good research should be creative and valid in the longer term. Some of the important features of good research are:

- Research purpose should be clearly defined
- Procedure for the research should be detailed sufficiently which should help the other to continue the work by referencing our work
- Research work should be carefully planned to get the results in related to the specified objectives
- Reports should be created by a researcher stating that what was the procedure adopted for completing the work which should also include errors in their findings

- Conclusions should be confident to those justified by the data of research

Research is guided by the rules of logical reasoning. The logical process of induction and deduction are of great value in carrying out research. In fact, logical reasoning makes research more meaningful in the context of decision making. Good research is empirical when the research is related basically to one or more aspects of a real situation and deals with concrete data that provides a basis for external validity to research results. Good research is recreated when research results are verified by studying the results and comparing the obtained results based on the decision which is sound enough to justify the research.

The research work should either form foundation for further advancement in the domain, draw some concrete conclusions or it should be beneficial from the social, commercial, or educational point of view.

Good research is systematic and logical. The report should be well written and it should be published through refereed journal. As, publications, copyrights, and patents help to reach the research to all intended audience.

1.9 Summary

Research inculcates scientific, curious and inductive thinking. It is an important component of the development of nation and individual. Research has special significance in deciding government policies in economics, in solving various operational and planning problems of business and industry, and in seeking answers to various social problems.

Research opens different avenues in particular domain for the betterment of mankind and world. Research activity develops critical thinking about the problem, systematic examination, developing and testing new theories, and draw important meaningful conclusions.

Ability of in-depth analysis and understanding a topic helps researcher to explore in many domains around. Research also helps researcher to sharpen the dimensions of his/her own personality in many folds.

Further Reading

1. Clifford Woody, "A survey of educational research in 1923," *J. Educ. Res.*, vol. 9, no. 5, pp. 357–381, 1924.
2. Chakravanti Rajagopalachari Kothari, "Research Methodology: Methods and Techniques", New Age International Publishers, New Delhi 2004.
3. Ranjit Kumar, "Research Methodology- A Step-by-Step Guide for Beginners", 3rd ed., SAGE Publications India Pvt Ltd, New Delhi, 2011.
4. David V. Thiel, "Research Methods for Engineers", Cambridge University Press, Cambridge, UK, 2014.
5. Robert Jeffrey Sternberg, "Beyond IQ: A Triarchic Theory of Human Intelligence", Cambridge University Press, Cambridge, UK, 1985.

2
Literature Survey and Problem Statement

Dr. Manoj S. Nagmode

CONTENTS

2.1 Introduction 26
2.2 Conducting Background Research 27
    2.2.1 Need for Background Research Study 28
    2.2.2 Incorporating the Background Information 29
    2.2.3 Background Research Plan 29
    2.2.4 Primary and Secondary Sources of Background Research 30
2.3 Resources for Literature Survey 31
    2.3.1 Systematic Manual Search 31
    2.3.2 Snowball Method 31
    2.3.3 Online Information Access 31
        2.3.3.1 Encyclopaedias and Dictionaries 34
        2.3.3.2 Reputed Journals/Impact Factor Journals 35
    2.3.4 Use of Interviews, Focus Groups and Questionnaires 35
2.4 How to Read a Scientific Paper, White Paper, and Patent 38
    2.4.1 Scientific Paper 38
    2.4.2 White Paper 42
    2.4.3 Patent 43
2.5 Research Reports 46
    2.5.1 Types of Reports 47
    2.5.2 Components of a Research Report 47
    2.5.3 Writing the Methods/Procedures and Findings 48
    2.5.4 Writing the Conclusions 48
    2.5.5 Publishing 49
    2.5.6 Authorship 49
    2.5.7 How to Read Primary Research Article 50
2.6 Recording and Summarizing the Findings of Literature Survey 51
    2.6.1 Finding and Selecting Resources and Materials Needed for Research 52
    2.6.2 Searching the Quality Material 52
    2.6.3 Scanning, Accessing and Evaluating the Literature 53
    2.6.4 Developing the Contents of Literature Review 54
    2.6.5 Where to Place the Literature Review in the Theses? 55
    2.6.6 Sequencing/Organizing the Literature Review 55
    2.6.7 Writing Literature Review 55
    2.6.8 Literature Survey Table: Case Study 56
2.7 Formulation of the Problem Statement 56
    2.7.1 Definition and Selection of a Research Problem 57

    2.7.2   Clearly Defining the Problem 61
    2.7.3   Method Used in Defining the Problem 62
    2.7.4   Study of Available Literature for the Problem Statement. 63
    2.7.5   Generating Ideas through Discussions for the Formulation of
            Problem Statement. 63
2.8 Scope and Significance of the Research Problem. 64
    2.8.1   How to Define a Problem Statement 65
            2.8.1.1   Review the Problem Statements Based On .65
            2.8.1.2   Poor Way of Writing Problem Statements. .65
    2.8.2   Errors in Defining the Problem Statement 66
2.9 Summary 66
Further Reading. 67

> Do all the good you can, By all the means you can, In all the ways you can, In all the places you can, At all the times you can, To all the people you can, As long as ever you can.
>
> —John Wesley

Learning objectives of this chapter are to:

- Comprehend the fundamental concepts of literature survey and literature review
- Identify and scrutinize the resources available for literature survey
- Distinguish between scientific paper, white paper and patent
- Organize, compare and contrast the components and findings of reported research
- Define problem statement and frame the scope and objectives
- Plan the research activity

This chapter will enable the researcher to

- Understand the difference between literature survey and literature review
- Efficiently search and summarize the reported literature using various online and offline resources
- Learn and understand the scientific paper, white paper and patent
- Formulate the problem statement, aim and objectives
- Plan the research activities under the guidance of the experienced researcher

2.1 Introduction

Research is a mind-set. It requires calm, patient and controlled mind, hard work by heart, logic by brain and ability to tolerate hurdles by positive thinking to achieve the required outcome. The most important preliminary task to undertake a research study is to pursue the existing literature available in the field of interest. The survey of literature and review of literature is an integral part of the research to acquaint with the existing knowledge or material available in the area of interest. Literature survey means penetrating through the available literature in the selected area of research, whereas, literature review involves assessment and examination of the reported contributions/findings. Recording, summarizing and evaluating the existing findings and being able to provide expert judgment of the methods and finding is also called as literature review. [1]

When you are travelling from one city (source) to the other city (destination), consider two possible ways of finding your destination. In the first case one can go by railway. In this case, there is no need to know the route from the source to destination. Second way is to use your own car. In this case, you need to take help of geographical map (i.e. supervisor/guide in research). In Google maps, various routes are available such as low traffic route, fastest route, shortest route and so on. You need to plan your journey to reach to the destination. In the same way, when you are pursuing your research, you need to plan your journey of research, you need a help of your guide (like a Google map). Also, you need to devise your own plan to reach the destination.

Literature survey helps you to find out many answers while pursuing the research. It can be time consuming in initial stage. [2] But if literature survey is not done in initial stage, then you might get stuck up while pursuing the research. Some authors have also given the meaning of literature review as the critical study of literature survey.

A researcher cannot progress in the area of interest without the literature survey. A literature survey can be divided into three parts

- Searching of the literature in the area of interest
- Collection/acquisition of literature or data
- Critical study of literature/review of literature in the area of interest

Literature review helps you to widen the knowledge in the selected domain. It is very important to know the contributions of other researchers in the existing area or domain. Moreover, questions like, what challenges, which problems, what advantageous and disadvantageous of the existing systems, should trigger the thought process. These questions help to focus on finding answers to those limitations and accordingly you can define your own problem statement. [3]

Literature survey and literature review should go hand in hand. After searching the material available in the selected area of research, researcher has to read it, understand it, and note down the important points. Again, find the latest literature, read it, understand it, and note important points. This process of literature survey and literature review should be continuous.

2.2 Conducting Background Research

One of the important steps in devising a problem statement is the background study for it. Conducting background research is very important part of all types of research. Background study of any topic consists of explanation of the area of research and context of the problem to be studied and solved. It helps to arm yourself with general knowledge to more effectively narrow down your research topic and finding relevant source to pursue the research.

2.2.1 Need for Background Research Study

Background research is an important part to undertake the research in the area of interest. The study of background research is useful in order to know what approach or methodology the other researchers have used for their research study. Background research is necessary in order to know how to design and understand the research topic. When devising a background research plan, you have to ask yourself a number of questions Such as

- Why you need to pursue research in the selected area?
- How are you going to pursue the research?
- What are the constraints of pursuing the research?
- Which laboratory is to be used for experimentation?
- Is the topic latest or demanding?
- Which is the best method for the experimentation?
- What are the limitations of existing technique?
- Is the research topic feasible?
- Does the research/problem statement have substantial research component?
- Is there any social use for the selected topic?
- Does it harm the society?
- Is there any parallel research being pursued by another person in the selected area?

Background research helps you to understand the subject and the nature of your research topic. Based on the background research, we can ask help from guide, seniors and colleagues. Background research is a continuous process. It should continue throughout the research process. It helps you to draw the aim and objectives of the research to pursue. During the study of background research, a researcher has full flexibility to consider utilization of various information sources and decide which are likely to be most useful. As the background research is a continuous process, it helps you to be ready to modify the requirements as the search progresses.

There is a need to ask questions in order to get the correct information about the background research. You need to ask proper questions related to the area of research. One need to identify the relevant questions. Relevant questions are those questions that will help you to design and understand the research topic and its limitations. To identify the limitations of the related research topic is very important, as it decides the correct parameters to choose for the experimentation.

Identifying limitations of existing research topic is possible only through the proper background research. If the background research is not conducted thoroughly then, it is quite possible that other researcher has already reported the results. on the same topic it may happen that the other researcher might have given the results on the selected research topic. In such cases, the researcher does not have the idea about the results already obtained. Therefore, the background research should be completed thoroughly.

In some cases a question posed may not be relevant to the topic. At that time, you need an opinion of more experienced researchers/person. Experienced researchers are nothing but those researchers, who have worked on the selected research topic. They will guide you whether you are going in the proper direction or not. Otherwise, you will get stuck while pursuing your research. You can ask questions to the expert such as, "Which concept/book/paper should I study to understand the selected area of research?" You should be specific when asking your questions. The expert will get puzzled by some questions. If the questions you ask are generic, then it will leave experts pondering the answers.

Research is also based on learning from the experiences of other researchers. Experience is important in order to avoid the repetition of mistakes which the other researchers have made. It also avoids blundering around the mistakes of others. Research topic should also include the effects on the human life. Therefore, the background research is very important in order to identify a series of problems that can be tackled individually.

2.2.2 Incorporating the Background Information

The importance of incorporating background information into the introduction or at the start of the literature is to provide the researcher with the critical information about the topic being studied. It includes

- Introductory detail of the area of interest
- Methodology used for the topic to be studied
- Broad overview of the topic to be studied
- Name of the authors/researchers who are authorities in the area of interest
- Important happenings or events in the area
- Keywords and subject related to the vocabulary terms, which can be used to search the database
- Local, national and international status of the work
- References and bibliographies that can lead to additional search
- Explanation about the fundamental study conducted in the past
- The important historical dates and events of the major work carried out related to the topic
- The information why and in what way the research problem is important
- Social relevance and contribution

2.2.3 Background Research Plan

Planning of the research is an important part in order to streamline the research process ideally. But in practical case the research may not go as per the plan. It may require much more time than predicted for the accomplishment of the task. There is no time limit or no one will decide the time required for the research to complete. It is actually a timeless and limitless work.

Some simple steps, the researcher can follow in order to plan and manage the process of literature search. These steps include:

- Identifying the research query or research problem
- Plan your search related to research area
- Experimentation based on existing techniques or methods

- Evaluating the results of existing techniques or methods and recording the results
- Revise and review your research plan based on your results obtained
- Try to test your database, based on new methodology or techniques
- Compare results of existing techniques with the new technique
- Analyse and synthesize the results obtained

Identifying research problem or research query is a tedious job. First, you have to decide what you want to search. You have to start with the broad idea of the research domain you want to look at. Asking yourself some questions related to the broad area, which you have selected, will help to exactly pinpoint focus down the research query or problem you want to pursue for the research, At this stage browsing on the Internet can help you to clarify your thoughts and finding exactly what you want to research. For example, "What is the importance of video enhancement?" Such questions should be asked to clarify the doubts.

Along with the background research we need to study mathematical formulas or equations that will describe the results of your experimentation. You can also plan for the background research on the past results of the same type of experiments or inventions. Researchers have to devise their own plan to reach to the outcome of the research. Otherwise you will go around randomly, until you finally reach to the outcome. That is the need of background research plan. It helps you to pursue your research in a systematic and proper way with less hurdles. Hurdles are always there, but to overcome those hurdles, planning of research makes your journey simple and happy. Therefore to avoid getting lost, you need a background research plan.

2.2.4 Primary and Secondary Sources of Background Research

Primary and secondary resources are important in the background research.

(i) **Primary source**: It is a document, which is created while pursuing the research. It is written by a researcher who lived during the time period of his/her research and wrote about the findings about it later. Background research of primary sources is useful for the secondary source of background research. There are various types of primary sources available such as: newspaper articles, letters, raw data, diaries/memories (also known as a firsthand account), laws, photographs, posters, interview of persons working in the area, drawings, records of different institutes such as government organizations, reputed educational institutes, speeches of experts. These primary sources help to generate the information, which is important for background information. Primary source of research are generally based on experimentations carried out by other researchers, clinical trials, surveys, and qualitative studies.

(ii) **Secondary source**: The secondary source of research takes help of primary source for the explanation about the background research. It also expresses the content of the background research. Secondary sources use the evidences of primary source to back up assertions. Secondary sources are generally based on systematic reviews, analysis of reviews, guidelines by other researchers etc.

Primary and secondary sources of background research are given in Figure 2.1. The secondary source may become primary depending on how it is used.

Literature Survey and Problem Statement

Primary Sources	Secondary Sources
• Experiments • Clinicaltrials • Surveys • Qualitative studies	• Overviews based on reviews, met analysis • Guidelines by other researchers • Decision Analysis • Economic Analysis

FIGURE 2.1
Primary and Secondary Sources of Background Research

2.3 Resources for Literature Survey

There are various resources for literature survey as shown in Figure 2.2.

2.3.1 Systematic Manual Search

Systematic manual search is done through the hard copy of research journal, paper and so on. A systematic manual search is time-consuming, but it is required so that one can understand the flow of research paper and how abstracting and indexing is used.

2.3.2 Snowball Method

In this approach, it is necessary to first identify recent papers, that is, primary sources. These recent papers are used for tracking back from the references of the recent paper. Also, cross references can be used for earlier publications.

2.3.3 Online Information Access

Online information access method includes the periodical visit to web pages of journals in order to know the contents of recent issues. Study material is available in various forms. One of the oldest ways to conduct a literature review is through books and materials published in magazines that are available in libraries in the form of hard

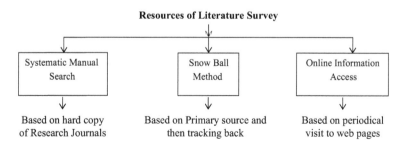

FIGURE 2.2
Resources of Literature Survey

copies. In the current era when the Internet is available almost in every part of the world, online resources plays an important role in literature study. A huge amount of information is available on the World Wide Web and one can identify research domains, available solutions, problems faced in current scenarios, and so on.

The problem faced in online resources is authenticity of resources available. In an online literature survey it is the responsibility of the researcher to use only that information or reference that is acquired from an authentic source. After choosing a topic, it is required to identify preliminary sources that can give basic information about the research topic. In case of unfamiliar or less researched topics, it is very important to find basic information from authentic resources at the beginning. Online resources are depicted in Figure 2.3.

Some online resources include:

(i) **Google scholar**: This is the best option. Google scholar contains scholarly research. But it also provides access to a limited number of journals. To use Google scholar, you have to put your search term and also put the time period on the left to get a good overview of a topic for search.

(ii) **Google Books**: Google books provide the digitized holding of the world's greatest academicians. It gives the world's greatest academic libraries. The sufficient background information to start your paper is available through Google books. It saves the time to visit to the library and to issue a book.

(iii) **Wikipedia (Evaluating e-books, journal papers, and websites)**: This is the free encyclopaedia that anyone can edit.

All of the online resources used for literature survey should be evaluated using some parameters.

(i) **Evaluating e-Books**: Any book used for background research can be evaluated based on
- Authority: Information about the author and publisher of the book
- Addressees: Information about the book, if book is written for some specific audience in some specific context or general
- Correctness: Has author consulted any other authorities to find accurateness of results?

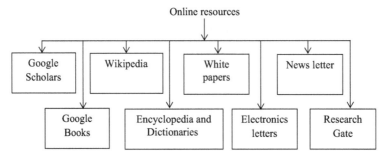

FIGURE 2.3
Online Resources

Literature Survey and Problem Statement

- Neutrality: Is book is unbiased and discusses all the possibilities of research problems?
- Relevance on time: Is book new or old? If it is old then is it still relevant in the current scenario?

(ii) **Evaluating Journal Papers**: Journal papers are usually evaluated based on their impact factor and process of review. Papers in reputed journal are rigorously reviewed by experts. Generally, such papers give more authentic information than articles or informative websites. The following things should be taken into consideration before using any journal article for reference:

- **Source of the journal/article**: You have to look for the article from scholarly journals which are written by the experts in research area. Some references may be available, which can lead you to the additional references of article or books. For some research topic a huge database may be available. In that case you can limit your search by the type of research article, review process, an editorial or a clinical trial
- **Length of the article/journal**: Note the length of the article. It can be a good clue to check whether the article will be useful for initial research or not
- **Authority**: Use authoritative sources in the research article. You have to use article which are written by experts in the research domain. The experts should be from the research domain and academic/research institute
- **Date of publication of research**: We need the latest information in the research domain. Research in almost any area requires the current information available. You need to ask yourself the questions such as: Is the article/journal you are referring is up to date for your research purpose?
- **Usefulness of the article/journal**: Is the article/journal you are referring is relevant to your research topic?
- **Audience referring the article/journal**: What type of readers are referring the article/journal? If the article/journal you are referring is written for some other research domain/area, then it will use the term/language/keywords special to that research domain
- **Citations**: Citation refers to the work published in past years. It is a reference to the previous case, used as guidance. Citation means you have to add a reference to the source where the mentioned idea/fact/statistics has been given. It is very important to cite the credible sources in your essay
- **Citation search engine**: Citation search engine are used to find the citations of scientific research paper. Different Databases are available to find citations of research paper or author. These include web of science (WOS), SCOPUS, Google scholar and IN-Recs
- **(WOS)Web of science**: This is one of the best tools for finding the number of citations of a specific article. It provides various options for finding citation. "Times Cited" section gives the number of citations in WOS
- **SCOPUS**: This gives a list of results including the number of citations that papers have received. SCOPUS is the largest citation and abstract database of peer-reviewed literature. SCOPUS indexing includes scientific journals, books and conference proceedings. Scopus h-index is used to measure the published work of scholar or scientist. It measures both the impact and

productivity of published work of a scholar or scientist. H-index is not a static value; it is calculated live on a set of results each time you look it up.
- **Google Scholar**: This search provides data on citations of authors and their published work. It searches for journals, books and chapters in the book. Google scholar shows information on the number of citations a paper has received among the different documents available in its database
- **Indexed Journals**: The quality of journals considered by the indexation of the journal. Indexed journals are of higher scientific quality as compared to non-indexed journals. There are various indexation services developed such as SCOPUS, EBSCO, PubMed, EMBASE, MedLine and SCIRUS

(iii) **Evaluating websites**: Authenticity of data available on various web sites is many times questionable. In this era of the World Wide Web, information about anything and everything is available on one click in the form of websites, blogs, social media pages, and so on. Checking the authenticity of such material is very important before using it for research. Anybody can add to such information without having actual knowledge of topic. To check whether information is genuine or not, check who are the authors providing the information? If the author is working in the same domain then information can be considered as authentic. Also check the references and bibliography given by author.

Other than authenticity, another concern in using online resources is copyright of information. Many times the information, code and data set available on the Internet is under copyright. Researchers should take care to obtain the required permissions before using any such information. Lastly, all material used for research should be properly cited in the documentation related to research.

2.3.3.1 Encyclopaedias and Dictionaries

Encyclopedias and dictionaries gives you overall views of the selected research topic. It gives keywords and hints for searching. List of sources are also available through encyclopedias and dictionaries.

Electronic database: We can search more than thousands of journals back a century or more through the electronic database.

Conference papers: Researchers should attend the conferences relevant to the area of research domain. In conferences, research scholar and professionals explore the latest trends, new ideas and presents new research. Finding or searching conference proceedings, one can get a feel of what is going on in a particular institution or within a particular research domain. Most of the electronic database include conference proceedings.

Internet: It is valuable resource of getting most of the information. We need to critically evaluate the information from the internet source. Research GATE is the best example of the collaborative scientific community that indexes the article available on the internet. We can find the full text of the article at no extra charge using the Internet So, at an initial stage we need to utilize the Internet for getting maximum information relevant to the topic and then critically evaluate the internet information.

Library: Library is available in all institutions/organizations. The libraries are well equipped with different varieties of resources. If you need resources which are not available in the library, then they can get these resources through interlibrary loan.

2.3.3.2 Reputed Journals/Impact Factor Journals

Impact factor is the most important measure to measure the quality of journal. Journal impact factors are based on journal citation reports (JCR). Journal citation reports are the unique data base which is used to determine the relative importance of journals within their subject categories. Impact factor of a journal is calculated by the number of citations received by the journal, from other journals within the (WOS) web of science database. The Thomson Reuters (Now Clarivate Analytics) impact factor is the most important impact factor.

Following are also the most commonly used resources for literature survey

- **IEEE electronic library**: It is basically for electrical and electronics engineering researchers
- **ASCE (American Society of Civil engineers)**: It is for civil engineering researcher
- **ASME (American Society of Mechanical Engineering)**: It is for mechanical engineering researcher
- **Nature**: Nature is a weekly International Journal publishing the finest peer-reviewed research. It is in all fields of science and technology on the basis of its originality, importance, interdisciplinary interest, timeliness, accessibility, elegance and surprising conclusions
- **Springer**: This is an American publishing company of academic journals and books
- **IEEE Explore**: This provides full text access to IEEE transactions, journals, magazines and conference proceedings. It is published since 1988
- **Science Direct**: This is the worlds best resource for research journals, abstract databases and reference
- **SciFinder scholar**: This provides the most accurate and comprehensive chemical and related scientific information
- **Math SciNet**: This includes world mathematical literature since 1940
- **JCCC**: J-gate Custom Content for Consortia: JCCC is a customized solution for sharing and accessing journal literature, which is subscribed by IITs, IISc and IIMs
- **JCCC@JNDEST**: It is a common gateway for accessing e-journals from different publishers subscribed by IITs, IISc and IIMs
- **ACM**: Association for Computing Machinery
- **SCOPUS: This is the Indexed Journal most commonly used for literature survey**

2.3.4 Use of Interviews, Focus Groups and Questionnaires

Interviews, focus groups and questionnaires are the primary research tools. They are basically used to collect original or new information through questions. Using these primary research tools, the researchers can frame the questions, specific to the research topic to be studied. They are used to collect database for in-depth understanding of the research project.

Interviews and focus groups are for the qualitative analysis of the research work to be carried out. These are based on discussions and some set of questions. It may include some visual observations and visual concepts. Most of the time it includes trial and error method. People opinions and their views are taken into consideration. If it is engineering research, the question may be what are the improvements the people want from the research?

Data collection is very important part of any type of research. Data collection is achieved through Interviews, focus groups and questionnaires. Some research problems need collection of some specific data from group of society or individual. Method employed by researchers for collecting primary data depends on type of problem as well as understanding of individuals from whom data is to be collected.

(i) Interviews

Research interview explores the outlook, understanding, and perspective of individuals on problem under research. Interviews are a qualitative method which provides a profound understanding of problem. They are predominantly suitable when the people may not want to discuss about the issues in a group environment.

Some important points to be considered while conducting the interviews:

- Proper/relevant questions to be framed
- There should be time limit for conducting interviews
- Opinions and feelings of the interviewee are important
- Recording of interview from the respondent, where appropriate
- For sensitive/social issues, obtain permission to record the interview from the respondent
- Collect the demographic information of the interviewee

Three fundamental types of research interviews are given in Figure 2.4.

Types of Interviews

In a **structured interview**, a list of predetermined questions is asked to all the individuals and their responses are noted down. No variation or follow up questions are allowed in this format. It makes the analysis of collected answers simple. This type of interviews can generate a database, which can be quantified easily and used for further statistical analysis. Drawback of this type of interviews is that the depth is not achieved in answers as individual is not allowed to explore the questions. For example: Asking questions to a class of students and noting their answers.

Unstructured interviews are more flexible and do not reflect any fixed type of questions, theories or ideas. The questions are not organized and flow of questions

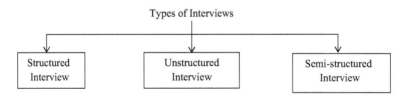

FIGURE 2.4
Types of Interviews

depends on response received from individual. The researcher needs to be more experienced to draw appropriate inference from the answers received. Unstructured interviews are time consuming and might fail to generate data, which can be quantified for further analysis. This type of interview is mostly used when depth is required in answers to analyse the problem further or area of the problem is less researched. For example: Asking questions in meetings or seminars.

The third type of interview is a **semistructured interview**, which tries to combine advantages of both structured and unstructured types of interviews. Semi structured interviews have list of some key questions but allow the interviewer to explore more depending on the previous answer and knowledge of interviewee. This helps to generate a dataset which can be quantified and at the same time gives depth to data. Also some times this approach helps to acquire data that is not previously thought of. Collective views of a group of a people is used for generating information, participant's experiences and beliefs are very useful in generating the information necessary for data.

(ii) Focus groups

A focus group is more than the structured data collection from multiple individuals in one place. A focus group is a facilitated group interview with individuals that have something in common. In this, combined information and various perspectives about some specific problems can be gathered.

To get the best results from the focus group, forming of correct group should be done with utmost care. The number of members of the group is one of the key factors. Very less number can generate incorrect data and very large number can generate chaos. There is no specific rule for how many members should be there in one group. It is up to the researcher's judgment to select number of members. If the group is non-homogeneous then perspectives and opinions of members can be too much varied which might not help researcher to reach to any conclusion. On other hand, if there is no diversity in group then there won't be much data generated as discussion on topic will be less. Interaction between members is a key to a successful focus group. If members are from diverse backgrounds then they might not interact comfortably. To avoid this many times groups are formed before the time, in which all the members already know each other and used for Focus group data collection. Conversely, in few specific problems, where discussion is not possible in known group because of society or any other pressure, more discussion and data is generated through group of totally unknown members.

For focus groups, there are no specific rules and it is up to the researcher to decide number of members and type of members. With properly formed group, focus groups can work successfully and generate useful data for research.

If it is a management research the questions may be what are the important people want to see in the business? Interviews and focus groups, also helps you to collect variety of data. As the data collection is a very important part of research. By asking different types of questions to different peoples, we can collect a variety of data. This data may be useful for qualitative analysis of research, which include textual or visual analysis observation, and so on.

One example of a focus group is a meeting of parents, which may include the views of the parents about the school or the institute.

(iii) Questionnaire

Questions are framed to get the answer from a group of people. Questionnaires are generally a fixed set of questions. These questions are used to collect the information for analysis. The questions are like the questions such as, Are you aware of the product available in the market? What are the weaknesses and strengths of the specific technique/method? and so on.

Questions are framed to be generally of the following types:

- Yes/no or true/false or correct/wrong
- Multiple choice
- Rating or ranking (i.e. agree or disagree or 1,2,3,4)

Before framing the questions, you need to test your questions with the help of a friend or family. Ensure that the questions are easy to answer and well framed. Questions should be short and relevant. Questions should be framed in proper word, which can be easy to understand. The questions you should like to frame, shouldn't lose interest of participants, for giving misleading answers. Questionnaires are usually effective, when carried out one-to-one. It can also be completed over the phone.

To have productive conversations with people, we need to prepare a list of questions. At initial stage, key points based on the research topic are to be prepared and from these key points, questions are framed. One can also frame subquestions related to some important types of the research topic. We need to make sure the answers must be recorded or take written notes or at least note down some important points related to questions.

2.4 How to Read a Scientific Paper, White Paper, and Patent

Scientific papers white papers and patents are important repositories for significant information. They provide valuable pointers and are most important to pursue research but reading and understanding the technical contents is challenging.

2.4.1 Scientific Paper

Scientific papers are most looked upon by the research community for understanding the concepts and contributions. However, it is difficult to understand scientific research paper at once. Some scholars are reading scientific papers as a text book. They are reading the paper from title to the end. They are digesting the contents, keywords and every word without any type of criticism. A scientific paper gives the idea about the Science/Engineering behind the research topic. It helps the reader to understand the logic/methodology of the research subject. Therefore there is a need to read a scientific paper carefully and with full attention.

Each part of the scientific paper serves a unique purpose and helps researcher to understand the research work correctly. The researchers, pursuing research in engineering or science community, scientific papers are the heart of science community. In a scientific paper, the results and ideas about new techniques/methods to follow are given. A scientific paper helps to understand the following things:

- What has already been discovered in research domain?
- What questions remain unanswered?
- How the experiments are conducted?
- What is the time duration to conduct that experiment?
- Which tools/equipment's needed to conduct the experiment?
- Details about how to perform the experiments?

All of this information is necessary to perform your own experiment. Sometimes, due to equipment constraints, it is not possible to conduct experimentation. At that time, the research has to find out whether the research is feasible or not. In such cases, the researcher has to apply for funding from industry.

In scientific research paper, the research article is divided into different sections such as:

- title
- abstract
- introduction
- literature survey
- methodology
- block schematic
- hardware and software used
- experimentation
- results
- discussions and conclusion
- future scope
- references

These sections serve a unique purpose, which help researcher to understand research work correctly. Now we will see how to read these sections one by one.

(i) Reading Abstract

The summary of the paper is given in the abstract. At initial stage, reading of the abstract gives you the overall idea of the research paper. After reading abstract, one can decide whether the research paper is in the selected domain or not. If the research domain matches with the abstract, then may one can read the other details of paper.

Abstract highlights the main problems of authors investigated as a research domain. It provides the key points or important results and observations from the experimentation. After that it gives the overview of authors' conclusion and discussions. Abstracts are generally available for free. They can be available at journals website or in the literature database. The abstract plays an important role, in deciding the overall contents of the research paper.

(ii) Reading Introduction

Background information about research paper is given in the introduction part of the research paper. It sets the background for the research work to be carried out. The introduction part is based on the citations and can be used by the researcher for additional background reading. The introduction gives the basic idea and significance of the topic.

(iii) Reading Literature Survey and Noting Some Important Points

A literature survey gives the idea about the research work about the other authors/researchers have already carried out.

(iv) Reading Methodology Used in the Research Paper

The methodology section gives the technical details of how the experimentation was carried out. It also talks about data set used for experimentation. This section gives actual technical details of work carried out by authors. It is very helpful to understand this section correctly for taking the research in same domain ahead.

(v) Understanding the Block Schematic of the System

The pictorial view of the technique/methodology used is given by the block schematic. It helps to understand the methodology used in the paper in an easy way.

(vi) Knowing Hardware and Software Used in the Paper

The types of hardware tools/kits used in the experimentation is given in this section. The results will vary as the hardware changes. Therefore, mentioning the specifications of Hardware is very important.

Similarly, what type of software tools used for simulation or experimentation is very important. The researcher has to specify the tools used for experimentation. As the hardware and software changes, accordingly to that the experimentation results will also change. If the experimentation is based on the performance analysis and the researchers are using different specification of hardware, then the result will vary.

(vii) Experimentation Done in the Research Paper

How the experimentation is carried out is given in the experimentation section. The experimental or surrounding conditions to be considered are also important, while performing the experimentation. The sequence of steps used in the experimentation is also very important.

(viii) Checking Results in the Scientific Research Paper

The results obtained after the experimentation are given in the result section. In this, the results can be given in tabular or graphical formats. In some scientific research papers, the results are given in various ways such as:

Literature Survey and Problem Statement 41

- quantitative analysis
- qualitative analysis
- time performance analysis

From this analysis of results, different types of information can be extracted such as how to represent the similar data for the comparison with different techniques/methods. One can also draw the conclusion from the results about the advantages or drawbacks of the existing methods. The researcher's contribution toward the research is given in the results. The research findings are reflected in the results. One can use these results for the comparison with his/her own findings or to use to build his/her own hypothesis. To get the most out of the results section, understanding of the graphs, pictures, and tables are necessary.

(ix) Understanding and Noting Down Discussion and Conclusion from the Scientific Research Paper

In the discussion section, the author explains about how the results are obtained. Also, the results obtained by the author are compared with other results and the comparison is discussed. Actually, the interpretation of results are given in the discussion section. The discussion is useful for comparing the results obtained from the other methods. Here the exchange of ideas between researchers takes place.

In discussion, the author can also highlight the existing weaknesses for the improvement. After the discussion, it is also observed that some questions are still unanswered.

The conclusion is based on the results. The actual results obtained are given in the conclusion.

(x) Knowing the Future Scope for Further Research

Based on the discussion and conclusion points, there are always chances of improvement, proposed by the researcher. These improvement raises the challenges for the further research study.

(xi) Scanning the References for the Past Work Done

The author has referred the information from other papers. These citations are listed in reference section. These citations can be used by researchers for additional background reading. Introduction gives basic idea and significance of the topic. Throughout the article, the authors refer to information from other papers. These citations are all listed in the references section. Both review articles and primary research articles, as well as books or other relevant sources, can be found in the reference section. Regardless of the type of source, there will always be enough information (authors, title, journal name, publication date, etc.) for the reader to find the source at a library or online. This makes the reference section incredibly useful for broadening your own literature search. If you're reading a paragraph in the current paper and want more information on the content, you should always try to find and read the articles cited in that paragraph.

2.4.2 White Paper

White papers are useful for R&D engineers, designers, finance, sales persons, and marketing persons. There are various definitions of white papers. Some of the definitions are given here.

- "A white paper is an authoritative report or guide helping readers understand an issue, solve a problem and make a decision." White papers are used in the main spheres, government, and business-to-business marketing. They may be considered as grey literature
- "A white paper is an article that states a company's position or philosophy about a social, political or other subject or a not too detailed technical explanation of an architecture, framework or product technology," posted by Magret Rouse
- "An information document issued by a company to promote or highlight the features of a solution, product or service. White papers are sales and marketing documents used to entice or persuade potential customers to learn more about or purchase a particular product, service, technology or methodology"
- "White paper is a value added content that can add value to your deliverables and your career" by Mak Pandit

The information in the white paper should be objective and truthful. Some important points need to be considered for writing white papers are

- Include accurate information from your research
- Avoid using biased or noncredible sources
- Include all relevant information
- A white should be educational, not promotional in tone
- It must attract the right audience: such as researchers, academician, and so on
- It must engage the reader or researcher
- The title is very important for white paper. It is a key for the success of the paper. Choose a title that conveys both the specific purpose of the White Paper and one that will gain the attention of your intended audience

White papers are generally written by institutions such as:

- Academic, Research and Development
- Government Organizations
- Company

White papers contain the following information

- Technical data, technical and facts
- Information about the process
- Comparison

Literature Survey and Problem Statement

- Analysis
- Opinions
- Conclusions
- Recommendations

2.4.3 Patent

Understanding patent specifications is an important skill for any researcher or technology manager. This primer gives a basic overview of a complex and constantly evolving topic to enable you to read and make sense of patent specifications relevant to your field of technology.

A **patent** is defined as a set of exclusive rights granted by a sovereign state or intergovernmental organization. It is granted to an inventor or assignee for a limited period of time in exchange for detailed public disclosure of an invention. An invention is a novel solution to a specific technological problem and is a process or a product. Patents are a form of intellectual property (IP). The details are:

(i) **Patent application**: For the grant of a patent, a patent application is filed. A patent application is a submitted to the patent office. It is actually a request pending in the patent office for granting the patent for the invention described and claimed by that application. It is also used to refer to the process of applying for a patent, or to the patent specification itself

(ii) **Patent specification**: The description of an invention is called as a Patent specification. It is submitted together with the official forms and correspondence relating to the application. It is a document describing the invention for which the patent is filed and setting out the scope of the protection of the patent. A specification generally contains:
- A title for the application
- Detailing of the background and overview of the invention
- A description of the invention
- Embodiments of the inventions and claims
- It may include figures to aid the description of the invention

Patent specification should comply with the laws of the office concerned. A patent is granted for the invention described and claimed by the specification.

(iii) Patent Prosecution

It is a process of "negotiating" or "arguing" with a patent office for the grant of a patent. It is also the interaction with a patent office with regard to a patent after its grant.

(iv) Patent litigation

Patent litigation is related to the legal proceedings for infringement of a patent after it is granted.

(v) Claims

A granted patent application includes one or more claims that defines the invention. A patent may include many claims. Claims define a specific property right. The claims must meet relevant requirements for claiming the patent such as novelty, usefulness, and non-obviousness. To prevent the misuse of the invention an exclusive right is granted to a patentee in most countries. It is the right to prevent others, or at least to try to prevent others, from commercially making, using, selling, importing, or distributing a patented invention without the legal permission. Claims describe the invention in a legal style. It sets out the essential features of the invention in a manner to clearly define what would infringe the patent. Amendment of claims are done during the prosecution to narrow or expand their scope. The rights of a patent for an invention lies with the first person to file an application for the protection of that invention. Therefore, it is beneficial to file the patent application as soon as possible.

(vi) Patent pending

It is a warning that an alleged invention is the subject of a patent application. This term is used to mark products containing the invention to alert a third party to the fact that the third party may be infringing a patent if the product is copied after the patent is granted.

(vii) Search

After filing the patent, a search is carried out, either systematically or in some jurisdictions. The main purpose of this search is to reveal the prior art which may be relevant to the patentability of the alleged invention. The patent is published 18 months after the priority date of the application, and as such is a public document. The search report is important to check, if there is a prior art which prevents the grant of the useful patent. In this case the application may be abandoned before the applicant incurs further expenses. Competitors can also use this search report, to get the idea for the scope of protection.

(viii) Examination

This is a process to ensure that an application complies with the requirements of the relevant patent laws. Examination is an iterative process. The patent office notifies the applicant about its objection. In such a case, the applicant may respond with an amendment or an argument to overcome the objection. The amendment or argument may be accepted or rejected. After that again the applicant waits for the further response and so forth until a patent is issued or the application is abandoned.

(ix) There are three types of patent applications.

- National
- Regional
- International

(i) **National applications**: National applications are generally filed at a national patent office. In the United Kingdom there is an office called the "United Kingdom Patent Office." It is used to obtain a patent in the country of that office.

(ii) **Regional applications**: This is an application that may have an effect in a range of countries. For example, The European Patent Office (EPO) is a regional patent office. It grants patents that can take effect in some or all countries contracting to the European Patent Convention (EPC), following a single application process.

Filing and prosecuting an application at a regional granting office is advantageous. It allows patents in a number of countries to be obtained without having to prosecute applications in all of those countries. The cost and complexity of obtaining protection is therefore reduced.

(iii) **International applications**: The World Intellectual Property Organization (WIPO) operates the Patent Cooperation Treaty (PCT). It is a centralized application process. The PCT system enables an applicant to file a single patent application in a single language. The application, called an international application, can, at a later date, lead to the grant of a patent in any of the states contracting to the PCT. WIPO, or more precisely the International Bureau of WIPO, performs many of the formalities of a patent application in a centralised manner, therefore avoiding the need to repeat the steps in all countries in which a patent may ultimately be granted.

(x) A general patent format is having following points.

Similar to a journal research paper, a patent can be read by understanding the following parts.

- Abstract
- Background
- Summary of the invention
- Detailed description of the invention
- Claims
- Figures

(i) **Abstract**: The abstract summarizes the invention. It is mainly used for patent searching purposes and is published with the application's bibliographical information on the respective office's patent search website.

(ii) **Background**: The background outlines the basic area of the invention. It also describes the work done in area before the patent filing date. The background is referred to as the "prior art." Typically the background also notes the shortcomings in current solutions. This gives the significance of the new invention to solve the problems identified.

(iii) **Summary of the invention**: This section includes statements that set out the key features of the invention which the inventor believes are novel and inventive over the prior art. Each different embodiment (example) of the invention should be identified in this section. If the invention is a combination of steps of a method, or features in a particular configuration, these combinations/configurations should be specifically defined. The statements in this section typically mirror the wording of the claims. Since the patent claims define exactly what must be done to infringe the patent, they must contain clear, unambiguous language. If there are words used in the summary of the invention that would be ambiguous when read by the skilled person, those words should be defined in the detailed description of the invention.

(iv) **Detailed description of the invention section**: The detailed description of the invention includes a detailed discussion of the features of the invention and how it would be made and used. It should also contain definitions of any terms used in the specification that may be ambiguous to a skilled person. The description often continues the narrative from the background. For example it may state how the invention solves the problems previously identified, or list benefits of the invention over the prior art. It is important to describe alternatives and variations of features so that a broader claim scope can be justified. For example, where functional groups could be substituted without affecting the core functionality of the compound as a repellent, these functional groups should be identified and details of optional substituents provided. Working examples of the invention are also typically included within the description. Examples demonstrate to the reader that the invention has been made and works as a mosquito repellent as promised. Patent examiners may only grant a patent if scientific evidence supporting the invention is presented in the specification.

(v) **Claims**: The claims are the key part of the specification because they define exactly what a person must do to infringe the patent. In this way, they define the scope of the patent itself. The claims are a list of numbered statements found after the description. They consist of a preamble and one or more features which outline the novel and inventive features of the invention. For chemical compound inventions it is allowable to include the chemical structure in the claim itself and define the functional groups of the formula.

2.5 Research Reports

An important goal of the research scientist is the publication of the results of a completed study. Scientific journals do not allow for literary embellishments and expressions, often seen in other journals, as the purpose is to communicate the scientific findings as clear as possible, in a highly stylized, distinctive fashion. This often makes it difficult for the applied professional to grasp all that the article has to offer. The purpose of this article is to help bridge much of that communication breach in scientific writing.

2.5.1 Types of Reports

There are many different formats for reporting research; journal articles, technical research reports, monographs or books, graduate theses or dissertations. Research is also reported orally at professional meetings, seminars, symposia, and workshops. These oral reports, however, are usually based on previous written reports

Journal articles are the most condensed form of writing. Journals have severe space limitations and often all the details of a complex research project can't be presented in one article. They are the most "prestigious" format for reporting disciplinary work. Involve a peer-review process which evaluates quality and importance of a paper. They receive wide distribution to disciplinary and subject-matter readers. To reach other audiences, other publications must be used. Sometimes, research can be written as journal articles as well as other formats to reach different users. Be careful, though, not to violate exclusive publication rights of journals – get permission!

Graduate theses and dissertations tend to be on the other extreme of length and completeness. This is the report of the student's work to his/her graduate committee These tend to be long and sometimes more wordy than necessary. But completeness is considered more important than efficiency in this writing. (In this respect, they are the opposite of journal articles.)

2.5.2 Components of a Research Report

The components of a research report often include: Title, Acknowledgments, Abstract, Table of Contents, Introduction, Literature Review, Conceptual Framework, Methods and Procedures, Results, Summary, and Conclusions.

- **Title**: This includes the name of the research report, Authors, affiliations, keyword and similar information written on title page
- **Acknowledgments**: Recognize the assistance and support of individuals and organizations
- **Abstract**: This is a compact summary of the research report, sometimes called "executive summary". The abstract is extremely important: It is the only thing that most people will read
- **Table of Contents**: A listing or outline of the organization of the report. It shows headings, subheadings and other divisions. Sometimes includes lists of Tables and Figures
- **Introductions**: Styles for introductions vary, from long and detailed to short or even absent
- **Review of Literature**: Serves the same purpose as in the research proposal
- **Methods and Procedures**: This section explains how the analysis portion of the research was conducted. It includes data collection and manipulation, data sources, analytical procedures, models developed and used, empirical procedures and techniques, and analyses conducted. Also includes problems encountered and how they were addressed
- **Results (Findings)**: Presents and explains the results of the analysis. This is the end product of all the analyses from which objectives were either achieved or not. Hypotheses have been tested and the results reported here

- **Summary and Conclusions**: This provided the reader with a general understanding of the research project. It most often includes an overview of the entire study, emphasizing problems, objectives, methods, procedures and results
- **Conclusions**: This represents the researcher's interpretations of the results
- **List of References**: A listing of all references used in every part of the report are included in this section
- **Appendices**: These can be very useful, but are not always used. They can be used to present material that might disrupt the flow of thoughts in the report (eg. too much detail) or include information of interest to only some readers. eg. Mathematical proofs or derivations, some statistical estimations or tests etc

2.5.3 Writing the Methods/Procedures and Findings

Writing these sections in a research report can have some important differences from the proposal. First, the methods are largely written for other economists. Others may be interested, but will tend to leave judgments of the validity of methods and procedures to the economists. The effectiveness of this section depends on the organization and thoroughness. The procedures need to be described in a logical sequence. Explain the data used, their source, and any manipulations or adjustments of the data. Explain and justify your analytical assumptions. Also explain models you may have used and be explicit about assumptions made. Describe problems you may have encountered and how they were resolved. Also, note unsuccessful approaches, techniques and procedures – this may help others to avoid problems or mistakes.

Be sure readers know how calculations were made and estimates derived. Define all variables, including units of measurement. (These details are easily overlooked, but are important.) Use graphs or other visual aids where appropriate to increase understanding by the reader; for example, a diagram of a multisector model may help to make linkages and equations used in the model more clear. This presents and explains the results of the study.

Validity of hypotheses are discussed, along with various test of validity used. Don't just present empirical estimates, these must be analysed, interpreted and possibly tested to make findings complete. The empirical results are often only the beginning of the most meaningful part of the research economic understanding, expertise and insight are needed to fully interpret the meaning and implications of the estimates.

Tables and figures are often effectively used to present findings. They help to organize and emphasize information in the findings. A recommended approach in writing this section is to construct the tables and figures that form the core of the findings first and then write the narrative which describes and explains the tables and figures.

2.5.4 Writing the Conclusions

Sometimes people confuse conclusions with results. Results are nothing but findings and they are used to test hypotheses. Conclusions are concerned with implications or tests of hypotheses. Conclusions address questions such as "so what"? Conclusions extrapolate beyond the findings: examining and interpreting the implications of the study.

Literature Survey and Problem Statement

Conclusions allow judgment of the researcher about implications of the study. This judgment must be supported by logic, but is subjective. Conclusions are a final inductive phase of the research, which are a matter of judgment. You may offer insights about the implications of your study and findings of other studies. Policy implications of your study may be considered, even if these were not among the objectives of the study. Conclusions may specify what the study implies as well as what it does not imply. Researchers may see a need to avoid improper use of the research results.

2.5.5 Publishing

Written research reports communicate knowledge within the research and scientific community. Publications are the primary means of disseminating research knowledge. Once published, research results become public knowledge. However, original ideas must always be recognized. In addition to journal publications, other outlets include, technical bulletins or reports, proceedings, papers, symposia, and workshop papers.

Among common reasons the proposed refereed publications fail to be accepted are:

- Inadequate identification of a research problem
- Inappropriate or unclear methods and procedures
- Inappropriate material for the proposed publication
- Failure to communicate what is important and original
- Poor organization.

2.5.6 Authorship

This can be a sensitive matter that deserves careful consideration.

Credit for the written research paper can be given in following three ways.

- Authorship of the research paper
- Citations or references
- Acknowledgments

The most prestigious type of credit given for the research paper contribution is the authorship. The researchers/persons, those who are involved "directly" and giving "important" or "substantial" contributions are included as authors. First author is generally assumed to have the greatest involvement in the research.

Interdisciplinary research may result in special difficulties in recognizing contribution and authorship. Research related to laboratory and field sciences, place special emphasis on the generation of data (sometimes as the end product of research). Economists are likely to view such data as "inputs" to their research, which only requires acknowledgement. Conversely, laboratory and field scientists may assume that activities (such as determining economic implications) are not part of "research" and thus don't deserve inclusion in authorship. Order of authorship can be a difficult issue.

The inclusion of administrators or advisors as authors, even if they had little to do with the research, can be a contentious issue. It is best to discuss authorship up front and openly among all involved in the research.

2.5.7 How to Read Primary Research Article

Abstract include the summary of the entire research paper. Generally experienced scientist avoids, reading the abstract at the start.

The following steps are used to read a primary research article.

- **Identification of important questions**

We need to identify important questions, in order to understand the research work carried out in the paper. The questions can include "What is the problem solved in this research?" "What are the methods/techniques used to solve the research problem?" These questions help you to identify why this research is being done. It also helps to find out the agenda for the motivated research.

- **Summarize the background in less sentences**

Summarization of the background in your own words is very important. This should include:

- What work has been done in this area of research to answer the important questions?
- What are the weaknesses/limitations of this work?
- What is the future scope, to overcome the limitations?
- Why this research is done/What is the motivation behind the research work?

This summarization points are very important in order to decide your research topic.

- **Study the Approach and Methodology Used in the Paper.**

The methodology used in the paper is the methodology followed for the experimentation. This methodology need to be studied for getting the correct results. We need to draw the diagram for each experiment. It shows exactly, what the authors did. We need to include maximum details in order to understand the work done.

- **Draw Conclusion from the Results Section**

Note down the results given by the authors. Whatever results given by the authors, we need to note down. Study those results and write some sentences in order to understand the meaning of results. In research articles, the results are summarized in figures, tables and graphs. We need to study those graphs and figures carefully. In most of the research articles, comparison of different techniques and results obtained are given in graphical form. It makes easier for the authors to represent their results. We need to draw our conclusion from those results. In some graphs, error bars are shown. These error bars indicate a lack of confidence intervals.

- **Study Dataset Used by Authors**

The results are obtained based on the database/dataset provided. The number of samples of dataset decides the effectiveness of the algorithm/methodology provided.

Complexity of the dataset decides the robustness of the algorithm. Study the dataset set in detail. What are the challenges involved in the dataset. It decides the execution of the algorithm. We can use the same dataset, as the test dataset for executing our own algorithm.

- **Carefully Read Discussion/Conclusion/Interpretation Section**

The opinions of the authors based on the results are given in the discussion and conclusion section. Whether the results obtained are correct or incorrect are justified in the discussion section. Any weaknesses/limitations identified in the research area are given in the conclusion section. In interpretation of results, the advantages and disadvantages are discussed and based on that the conclusion is drawn.

- **Now Read the Abstract**

After reading the whole paper, it is time to read the abstract. In the abstract, one need to check whatever the contents given in the paper matches with the abstract or not? Does the abstract also reflect in the conclusion section of the paper. List out the opinion of the other researchers about the paper. In order to continue your research in same domain, opinion of the research paper given by other researchers is important.

We need to identify the experts in this particular research field. These experts must have studied the research area. They should think critically about the research domain. The criticism given by expert researchers should be considered seriously. One can also use Google to find the possible criticism or challenges involved in the research domain. Then only, you will prepare your mind set in order to do the research in the selected area.

2.6 Recording and Summarizing the Findings of Literature Survey

In literature, a summary of the work done in existing research topic is given. Generally, it is all based on the secondary sources, that is, what other researchers/people have already written about research project. It is not concerned about the novel or new discoveries about the research. Actually, it should find the way for further research.

The central interest of literature survey is always on the theories put together by the recognized experts in the research domain/field. It also includes the data collected from the different experts, who have already worked on the same research topic. A literature review will try to find the existing research, which is already done in the selected research domain. It should review the important scholarly books in the relevant area. But, reading only scholarly books is not sufficient, one should take a keen interest in journals articles. Journal articles give you more updated information about the research subject.

Preparation of literature survey involves following points

- Summarizing the key points from the literature
- Searching of accurate, reliable, and latest information on the research topic or subject

- Listing the ideas and concepts into a summary of what is known
- Synthesizing, discussing and evaluating these ideas and concepts
- Criticising and identifying the particular area for debate or controversy
- Preparation of mind-set of the researcher and relevant resources for the application of these ideas for new research to be proposed

2.6.1 Finding and Selecting Resources and Materials Needed for Research

Availability of resources for selected research topic plays an important role. If the proper resources are not available, then you cannot conduct your research work. So, first thing is always to provide proper resources for the experimentation purpose.

We need to ask following questions such as:

- What is my research title or research focus?
- What kind of hardware and software tools, do I need to carry out experimentation?
- What are the different types of journals/books/government documents available for literature?
- Is my search wide enough to ensure that I have identified all the relevant material?
- On the contrary, is my search narrow enough to exclude irrelevant material?
- Is the literature, good enough for the Ph.D level?
- Have I considered the alternative or other points of view?
- Will the other reader find my literature review more relevant, appropriate, and useful?

2.6.2 Searching the Quality Material

The major thing, you have to look the most up-to-date or latest material for the research subject. In this only the important book/reference books, which are written by experienced experts and leading scholars should be taken into consideration. Good literature review sets out the many perspectives as possible. One has to balance between substantial academic books and the latest journal articles.

One can also look for the bibliography provided with the module documentation. Choose one or more similar looking contents of books or articles and then scan through the bibliographies provided by these authors.

Keywords are important for finding the literature from the library catalogue search engines. We can try number of other keywords to capture as many items as possible for the varieties of data. We also need to look for over generalizations. Try to be specific for some research related area. Afterwards, we can narrow the field down, so you get just a few dozen results. If the material collected through the search engine is not sufficient, then browse library shelves in the relevant subject area and look for the book, which can capture your eye. Check the index and contents of the book to see if they are likely to help. If not put the book back and try again to find the other relevant material for the research subject.

We can also take the help of librarian to get the proper book from the library supervisor. Research guide will also help you for getting the proper direction for the literature survey but remember, the survey must be done by the research scholar.

Literature Survey and Problem Statement

There is no limit for referring to journal articles. The number of journal articles, you need to refer depends on the following points:

- The type and nature of the subject
- Duration of the research work to be carried out
- Level of study etc

Find out the obvious gap in the existing literature survey. It helps to provide the scope for improvement. You have to use your judgment for referring to the number of books and research journals. But the references should be from varied sources and one should capture the latest information about the research subject taken into consideration. If your subject involves new emperical research, then you need the latest literature presented by the author. At the same time, we need to consult with experienced researchers in the same domain.

2.6.3 Scanning, Accessing and Evaluating the Literature

After identifying the relevant research article and books pertaining to your research subject, you need to scan through their research article, journal and books. Try to find out the idea or clue, which will help you to contribute to your study. If the researcher had got some clues, then he can concentrate on some selected research/journal article for further study. If the researcher didn't get some idea/clue from the literature, then he has to choose some different research article/book.

After identifying some clues the researcher has to read the selected/chosen material carefully. In the selected material one has to understand the concept and logic, given in the article and paper. One has to look especially for the following points:

- What are the keywords given by the author?
- What are the proofs and evidences the author has produced to support the idea presented in the paper?
- The evidences interpreted by the authors can these by represented in other way?
- What type of results provided by the authors, such as quantitative, qualitative or experimental?
- Is there enough mathematical analysis given by the author, to support the research work?
- Comparison of theoretical values with the practical values
- Critical evaluation of other literature
- Has the author included the literature, which has opposed his/her idea presented in the paper?
- Validity of research information provided in the paper, is it from the reliable source?
- Is it possible for you to provide your own judgment to deconstruct the argument?
- Is it possible to identify the gaps or future scope for improvement in the existing research?

- What type of information or contribution is provided to your own research subject?
- Are there any strengths and limitations given in the paper?

While scanning, accessing, and evaluating the research article, you need to note down the important points provided in each research article and book. Even at initial stages, you didn't find these notes important, you may need these notes for a later summary. So always keep these references with you after indexing them properly.

2.6.4 Developing the Contents of Literature Review

The literature review will contain a number of "mini reviews" based on the material read. Every mini review contains the following important points:

- A brief summary of the article/book
- Critically evaluated points in the book or article:

  (i) **Brief summary of the article/book**: along with the brief description of the research presented in the book for your own research subject. At least one key point or keyword is given in each research article or book. One has to understand this key point or keyword and the relation of this key point with your own research. This key point is related with new techniques/method/claim given in the research article book.

  (ii) **Critically evaluated points in the book/article**: Critically evaluated points include the strengths and weaknesses given by the research article/book. If the strengths and weaknesses are not provided by the author directly, then researcher has to identify the strength and weaknesses of the research article/book. After going through the conclusion and discussion sections of research article/paper. One has to find out the following points

  - Does the author produces sufficient evidence to establish the point they want to make?
  - The researcher has to identify the points that the author has conveniently left out or skated over
  - What is the author not saying related to the work carried out?
  - Has the author given proper answer to the arguments?
  - What are the evidences for those arguments?
  - Identify carefully the loop holes in the presentation
  - Generally the author is spinning the evidence. Is it possible for you to spin the evidence in another direction?
  - What is the impact would this have on overall argument given by the author?
  - Here you find some clues for research area. Based on your reading the material, you need to ask yourself, what if this particular author is completely right or partially right or completely wrong

If you are able to answer these questions, then it means that you have critically evaluated the research article or paper.

2.6.5 Where to Place the Literature Review in the Theses?

Generally the literature review is placed, immediately after the introduction part of the theses. This is the "traditional" way to place the literature review to be distributed over the theses as a whole. The advantage of this is that, the entire theses reads like a continuous and ongoing process throughout the research duration. The decision to place the literature review should be taken only after the discussion with your supervisor/guide.

2.6.6 Sequencing/Organizing the Literature Review

The structure and the sequence of the literature review section or chapter should have the beginning, middle and end. The chapter, based on the literature review could guide the reader based on the following points:

- Strategy you have adopted for selecting the books or article
- Presenting the topic theme for the review
- Analysis from the referred book or article thoroughly
- Conclusion of literature review should be based on the key point referred from article/book

One has to work systematically between the article/book selected. Make some subsections based on the research topic selected. A paragraph is generally enough for one article/book selected in the research paper.

For each article/book one has to commit on following points:

- Brief summary of the main idea
- Limitations and weaknesses of the existing subject
- Methodology followed by author
- Relevance of the book or article for your research subject
- Future scope given in the article/book

In a Ph.D theses, the literature review is typically one chapter. The content of this chapter will vary depending on the subject of research. Take help of your supervisor/Guide for the preparation of literature review. One "traditional" approach is to organize your literature review into historical orders. Some researchers prefer to organize the literature review with the main theories and giving more space to theory and keeping the less important stuff later. Some researchers prefer to group some key ideas, concept, fact and approaches together and then bouncing the idea of each other.

Whatever the approach is adopted by the researcher, one has to take care that, your review should flow smoothly, that is, one idea of book/article leads neatly to the next. Reader should be able to understand the literature review effortlessly through a sequence with a clear, precise, correct, accurate and interesting information.

2.6.7 Writing Literature Review

Writing literature review starts only after taking adequate thinking time. When you have done all the reading and planning for the different sections, then start writing. The introduction to the literature review includes mentioning of the references used for the research subject. Guide the reader through the material presented neatly and clearly. Do not make claims directly, it should be supported by evidences. If it is management related research, then there may be too many "Quotes." Support "Quotes" with the proper evidences. Avoid too many "Quotes."

The summary of the literature review should be done in your own words. Here are some important points to be noted for writing literature review:

- **Claims should be supported with evidence**: whatever the claim given in the literature review, it should be supported with the proper evidences. You should be careful while interpreting the evidence
- **Keywords from the literature**: From each literature survey, only selected points should be considered. Use your judgment to identify what is important and what is secondary from the literature
- **Summary from the literature**: Your own words should be used to summarize the finding from the literature
- **Arguments should be presented with evidence**: Use your own thinking and voice with a clear-cut argument. Arguments should be supported with evidence
- **Avoid too much I/we language**: It is better to avoid too much I/we language. Use more indirect language such as "It is observed from the results," "It could be concluded that," and so on
- **Revision of drafting**: Revise, refine, and edit the drafting, number of times. Grammar and spelling should be checked number of times. Fluency of language should be checked as well as the references that you have used

2.6.8 Literature Survey Table: Case Study

Examples of preparing survey tables for literature survey are given in Tables 2.1 and 2.2. In this table the research is based on object detection and classification. Column number three gives the limitations of the work carried out in the research area. Similarly Table 2.2 shows the comparative study of action recognition and behavior understanding methods.

2.7 Formulation of the Problem Statement

After finalizing the topic it is required to do background study of the topic to finalize the problem statement. Appropriate background study helps to make the problem statement precise and clear.

Every problem statement should have following characteristics:

- Known fact: Problem statement should be based on some already known fact which can be used to introduce the topic
- Gap: It should identify gap from the current solution
- Appealing: Interest of the Researcher should be preserved

Literature Survey and Problem Statement

TABLE 2.1

Literature Survey Table Based on Object Detection and Classification

Subtopic	Methodology used	Limitations
Object detection [4–9]	Methods to detect single and/or multiple object have been proposed using single or multiple background models	Multiple object detection in varying environments is a challenging task
Object Classification [10–11]	Semi-supervised and unsupervised classification methods using single or multiple features have been proposed. Shape based and motion based features such as area size, compactness, speed etc. are used for classifying objects	Classification of objects in complex environment and with low computational cost is still an unsolved problem
Object Tracking [12–15]	Region-based, contour-based, feature-based, model-based and hybrid approaches have been proposed.	Tracking of multiple objects and group tracking does not give accurate results
Action Recognition [16–19]	Direct and indirect recognition of activities quantizing feature vectors, shape mask, human kinematics and model based approaches have been proposed	Recognition of actions in cluttered scenarios is still a problem
Behaviour Analysis [20–23]	Supervised and unsupervised models using SVM classifier, optical flow based methods have been proposed	Complex action having interaction between two blobs are difficult to analyse

- Exploratory: can be investigated through the collection and analysis of data
- Noteworthy: It should contribute for the improvement in the research subject
- Feasible: It should fit the researcher's level of research skills, available resources, and time restrictions
- Ethical: Solution proposed should be ethical and should not harm anybody
- Challenging: Problem statement should be challenging
- Clarity: Problem statement should be clear cut defined

2.7.1 Definition and Selection of a Research Problem

Definition of Research Problem

A problem statement is actually a challenge based on the limitations, which a Researcher identifies in context of a practical or theoretical situation and wants to find a solution for the same. Problem statement formulation is really a challenge in the research domain. In any research, may be engineering, management or medical, the first and important step is to define the research problem in correct words. It should clearly define what type and improvement in the research is going to propose by the researcher. Researchers must examine all the possibilities concerning the problem before defining the research problem.

To define a research problem, a researcher must know the limitations/weaknesses in the selected research subject. The research problem can be defined for an individual or a group of persons. Technical description of a problem, which the individual or a group of people faces, is called as a research problem. If the individual or group, wants a desired outcome, then there are different ways (or methods) to achieve that outcome. These

TABLE 2.2

Comparative Study of Action Recognition and Behavior Understanding Methods

Sr. No	Ref. no	Name of the Author	Methodology Used	Limitations
1	[20]	Carolina Garate, Sofia Zaidenberg, et al.	Group tracking is done using seven features. Behavior recognition is done using knowledge model and event model	Prediction of dangerous events not done.
2	[24]	Tian Wang, Hichem Snoussi	Histogram of optical flow orientation is computed and nonlinear one class SVM classifier is applied	Small deviations in scene are not detected
3	[25]	Jasper R. Van Huis, Henri Bouma et al.	Track based secondary features are derived from basic features and rule based event classification is done	Only single track actions and interactions under controlled environments are used.
4	[26]	Maria Anderson, Fredrik Gustafsson et al.	Unsupervised Kmeans clustering and semisupervised hidden Markov model is used	Does not work under occlusion and for large crowds
5	[22]	Sofia Zaidenberg, Bernard Boulay et al.	Trajectories of people are analysed and clustered using mean-shift algorithm. An event description language is proposed	Internal group movements are not recognised
6	[27]	Xu, M., Zuo, L., Iyengar, S., Goldfain, A., DelloStritto, J	Semisupervised training approach is proposed for detecting simple and complex activities	Limited number of activities are considered
7	[28]	Vishwakarma, S., Agrawal, A.	A non-hierarchical approach is proposed. Feature vectors of interest points are quantized using a histogram	Computational complexity is more
8	[16]	Vishwakarma, S., Sapre, A., Agrawal	Multiclass activities is considered in a three dimensional coordinate system which is robust to scale and view changes	Number of features required are more
9	[17]	MahfuzulHaque, and ManzurMurshed	An approach based on shape masks obtained from background subtraction to represent actions is used	Objects smaller in size are not detected
10	[18]	Ronald Poppe et al.	A holistic approach using global body structure and dynamics is used to represent actions is used	Tested only on small set on actions
11	[29]	Yan Song et al.	Each depth pixel is mapped to corresponding RGB pixel. Spatial–temporal interest point (STIP)detector is used to generate initial points and KLT tracker to generate trajectories.	Method is specific for Depth cameras. For MSR_daily dataset proposed method outperforms other methods.
12	[23]	Nick C Tang et al.	Information from offline multimodal dataset is used to retrieve the corresponding depth map and skeleton structure.	Offline multimodal video dataset is required

methods may not have equal efficiency for the desired objective(s) and there is always a doubt which method (or course of action) is best.

Usually a research problem does exist for the following conditions:

- Individual or a group has some difficulty or the problem
- Objective(s)/outcome(s) to be attained. If there is no objective/outcome, one cannot have a research problem

- There must be alternative methods/means for achieving the objective(s). At least two methods must be available to a researcher for finding the solution
- Some doubt must remain in the mind of a researcher while selecting the alternatives. It means that the Researcher must answer some questions concerning the relative efficiency of the possible alternatives
- Based on the environmental conditions, the difficulty varies. With different environmental conditions, difficulty level also changes
- What type of input or database used to check the desired outcome

From above points, one can understand that a research problem is one which requires a researcher to perform number of experimentation in order to find out the best possible solution for the desired problem. It means that, the researcher has to find out by which methods (alternatives) the outcome/objective can be attained with optimum performance in the context of a given environment. Different factors need to be considered, which may result in making the problem more complicated. The number of alternatives may be very high. Persons not involved in making the decision may be affected by it and the person may react to it, favourably or unfavourably. All of these factors are considered in the context of a research problem. Researcher has to consider the challenges in the context of a research problem.

But still, we have a plan with us to pursue the research. From the data we have collected, from the literature survey, we need to plan different things, in order to identify important issues with the research work you have planned to do. One has to specify his/her own interest in the fields such as:

- Education: in terms of displaying video on LCD
- Advertising: in terms of creative designs, media channels, and so on
- Households: watching video on television, sending or receiving video through digital media

At this point, there is a need to find the limitations of your investigation. For finding the limitations one can consider the following parameters.

- Time: For example, processing time for the execution of the algorithm
- Internet BW: For receiving or transmitting the data or information
- Legal issues: To check the legal issues in processing the data or information
- Quality: Quality of the output
- Storage: Storage space for data

Based on these parameters, we need to focus at an early stage at some specific research area. After taking the basic review we need to again refocus on the questions later.

Plan your search according to the selected research area. There are a number of approaches:

- **Retrospective**: In the retrospective approach, one has to find out the most recent or latest method/technique and then work backward. In this the journal articles are referred to work backwards

- **Systematic approach**: In this approach, one has to find out all relevant material available in the library
- **Targeted approach**: In a targeted approach, one has to pinpoint the topic and concentrate on the narrow area of the research topic. One can target more on the selected area, when we have a clear picture of what we need to find out

Selecting the research problem is the difficult task. The researcher has to select research problem after discussing it with the research guide. But, at last, every researcher must select his own research problem. It depends on the mind-set and the passion of the researcher, to select the research problem. The Guide will only help the Researcher for the selection of Research subject. But the research problem has to be selected by the researcher only.

The following points should be considered by the researcher in selecting the research problem:

- Too narrow or too vague statements for the research problem should be avoided
- Researchers should avoid the selection of controversial subjects for the research problem
- Research subjects that are done by a number of researchers generally avoided for the selection of research problem
- The research problem should be feasible for experimentation/implementation
- Sufficient numbers of publications/materials should be available for finding the limitations/weaknesses of the subject
- The researcher's qualification, the importance of the subject, costs involved for experimentation and time factors should be considered

Even though these points are considered, it is somewhat difficult to propose definite ideas concerning how a researcher should obtain the research problem. Researchers should take advise from an expert in the same research domain. If the literature survey is done sufficiently, then it is easier for the researcher to discuss his/her own ideas with the expert in the same research domain. After discussing with the experts, the researcher has to obtain his/her own ideas for the research problem. Researcher can also discuss the research topic with his/her colleagues. In this way, the researcher has to make all possible efforts in selecting a Research problem. The researcher has to ask the following questions to himself before the selection of the final research problem:

- Whether sufficient resources are available for the researcher to carry out the research
- Budget provided/available for the research
- Cooperation from other colleagues/faculty to carry out the research work

After answering these questions, the practicability of the research problem is to be checked and the decision should be taken for the selection of the research problem. When the research problem is relatively new, and the sufficient well-equipped resources are not available, a brief feasibility study should be undertaken. This step is important because, this increases the confidence of the researcher to undertake relatively new research problem. It also makes the research interesting for the researcher and it will not

Literature Survey and Problem Statement

become a boring subject for the researcher. In other words, zeal and zest for the research work is very important. The research is not a short-term process, it is a long-term process, therefore careful selection of research problem, which increases the interest and confidence of the researcher is a must. The researcher has to prepare his/her mind-set for the research problem, then and only then the research will be the uppermost place in his/her mind. The researcher should undertake all types of efforts and pains needed for the research and to achieve the required outcome.

The sequence for the selection of problem statement is given in Figure 2.5.

In the selection of the research problem, first select your own field of specialization. You have to study critically the available research literature. There may be number of difficulties and gaps. Accept these difficulties as a challenge. Acquaint with recent research and trends of particular field. The researcher should be aware about the sources of problem and the development occurring in the technology and society. Get acquainted with the records of previous research. Discuss important points with the experts. Also attend seminars and exchange ideas with the other researcher. The researcher should have the questioning attitude. The research problem statement should highlight the interest of the problem statement. Check the feasibility of available resources for experimentation. Also check the availability of guidance in the selected research domain. The researcher should have the courage and confidence in the selection of research problem.

2.7.2 Clearly Defining the Problem

Clearly defined research problem help researchers to be on the right track, whereas an ill-defined research problem may create many difficulties and hurdles while pursuing the research work.

A clear and well-defined problem statement should bring the following questions:

- What are the limitations/weaknesses in the existing research subject?
- What is the database required?
- What are the characteristics of data, which are relevant and need to be studied?
- What parameters need to be explored?
- What are the exiting methods/techniques available for this purpose?

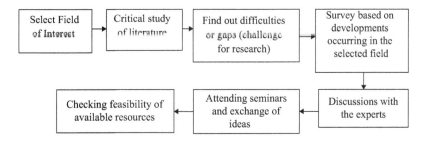

FIGURE 2.5
Sequence for the Selection of Problem Statement

Above questions, decides the planning of the research based on the clearly defined research problem. Clearly and properly defining the research problem is a step of the highest importance. Careful detailing the research problem occurs only when the researcher, puts all his/her efforts while pursuing the literature survey and that also comes out from his/her heart and mind. Formulation of a research problem is more important than its solution. It is due to the fact that the solution comes afterward, while pursuing the research work.

2.7.3 Method Used in Defining the Problem

The research is to define the problem within the boundaries within which it is to be studied. The research problem is defined with the predetermined objectives in mind. It is a complex task to define the research problem which must be tackled intelligently, to avoid the difficulties encountered in research operation.

Generally, the approach used to define the research problem is that the researcher should himself generate a question for the research subject. In some cases, someone else wants the researcher to carry on the research, it may be the concerned individual or an organization or an authority to pose a question to the researcher for defining the research problem. But this approach does not produce the definite results because the question phrased/posed in this fashion is generally in broad general terms and it may not be in a form suitable for testing. As discussed earlier, defining the research problem is a crucial part of the research study and it takes time for defining the research problem. One should think calmly before deciding each and every word in the research problem, after considering all the dimensions of the research problem. A systematic definition of problem is very important after considering weightages to all the related points.

Following points need to be considered for defining the research problem:

- Available literature need to be studied
- Nature of the problem need to be studied
- An idea is generated after the discussion with colleagues, friends, seniors and experts in the research subject
- The research problem need to be rephrased into a working proposition

For all the above points, a brief description needs to be written

Important steps for the method involved in defining the problem statement:

- Initially, the problem is written in a broad general way. This is done due to the practical feasibility or due to some intellectual/scientific interest
- The researcher should think thoroughly before posing the research problem
- If the research topic, is related to the social research, then, it is better to do some field observation. For this purpose, the researcher should do some preliminary survey. It is also called as the pilot survey
- Next, the researcher should take advice from the supervisor/guide or the subject expert in accomplishing this task. Guides help the researcher to put forth the problem in some general statement. But it is up to the researcher to narrow the problem and write the problem in operational term, to pinpoint the exact research domain

Literature Survey and Problem Statement 63

- If the researcher is working in some research institute or organization, then he can take help of the expert from the research institute or organization for stating the problem statement
- If the problem contains some ambiguities information, then it should be resolved by thinking over the problem again and again, with the cool and calm mind
- While rethinking about the problem, the feasibility about the particular solution has been considered. Feasibility and available resources should be taken in mind while stating the problem

In defining the problem, one has to understand its nature and origin clearly. For this purpose, we need to discuss it with the people, who first raised it. The discussion gives the idea about how the problem originally came and what objectives/outcomes expected from the research. Even though the problem is stated by the researcher himself, he has to consider all the possibilities concerning the problem statement. For understanding the nature of the problem, he must discuss the problem with the experts, who has a good knowledge of the problem concerned. The researcher has also to find out the other similar problems related to the research subject. The environment also plays very important role, in deciding the problem. The environment affects the problem to be studied in different situations.

2.7.4 Study of Available Literature for the Problem Statement

All available literature material, which is in hand with the researcher, must necessarily be surveyed and examined before the definition of the research problem. It means that, the researcher is conversant with the relevant theory concerning the subject to be studied. Also, records and findings related to the research topic is studied thoroughly by the researcher. There is always a need to devote sufficient time in revising/reviewing of the research, which has already undertaken by other researchers in related domain.

It also helps to identify, what type of data and material is available for operational purpose. The availability of data helps to narrow down the problem itself. It is also easier to decide the technique/method to be used for the experimentation purpose. Study of available literature survey also helps to identify the gaps in the theories or whether the existing theories applicable to the problem under study. All of these points help the researcher to think about new ideas, which occurs in his mind to take new strides in the field for improving the knowledge. Such type of study also helps the researchers to find out the new lines of approach to the present problem.

2.7.5 Generating Ideas through Discussions for the Formulation of Problem Statement

While pursuing any type of research, discussion with colleagues and experts gives you many ideas and produces useful information. Discussion with others is often known as Experience Survey. Other researchers with rich experience are in a position to provide maximum help in different aspects of the proposed study and their comments are very helpful for the researcher. Researcher has to take advice given by expert seriously. This will help sharpen the research problem statement and the focus of attention on specific aspects within the area of research subject. Researchers expect the points such as general approach to the given problem, techniques that might be used, possible outcomes or solutions and so on.

Rephrasing the problem into analytical or the operational terms is not a difficult task. The following points need to be observed while defining the research problem.

- Assumptions related to the research problem should be clearly stated
- Different technical words, terms and phrases with special meanings, used in the problem statement, should be clearly defined
- Criteria for the selection of problem should be provided
- Sources of data available and the time limit should be taken into consideration
- The environment/circumstances within which the problem is to be studied must be mentioned explicitly in defining a research problem

Defining the research problem the following steps are generally used

- Define problem in a broader sense
- Resolve the ambiguities involved
- Rethinking about the problem statement and redefining

Continuously thinking about the problem statement, result in a more specific formulation of a problem Statement: It becomes realistic, in terms of the available data and resources and it also becomes analytically meaningful. It results in well-defined problem statement and is capable of paving the way for the development of working hypotheses.

2.8 Scope and Significance of the Research Problem

It is not easy to clearly define the problem statement. There are some examples in scientific research, where the researcher might spend years for exploring, thinking and researching, before they are clear about what type of research questions they are going to answer. In most of the cases, when the research problem is too wide, the feasibility also needed to be checked. Also, the time required and the resources for the research problem are very important. The depth and focus of the research problem should be understood from the title. The problem statement helps you to keep on track with your research. The research problem statement should be adequate. Researcher generates many problem statements that may arise from the same situation. But in practical case, your research will pursue only one problem statement in depth.

The problem statement should have the following characteristics:

- The problem statement should be based on factual evidence
- The problem statement should be meaningful and testable
- It should be relevant and meaningful
- It should reflect the need for the society
- The problem statement should include research component
- The problem specified in the problem statement should be solvable, achievable and measurable

The problem statement should make it clear about the purpose of the research? The purpose of the research should be clear to you and the reader. The subsequent elaboration of method should be provided for the research.

2.8.1 How to Define a Problem Statement

The problem statement should highlight the limitation/negative points of the current situation and explain why this matter. The problem statement should be a great communication tool. It should help to get the support from other. The important goal of problem statement is to define the problem in a clear and precise way. The goal should be to focus onto the improvement of research activity. Creation of a problem statement, involves the following steps:

- Write different statements, related to the problem. Compare each statement, looking for common things and working
- After comparing different statements, write an improved statement related to the common theme
- Ensure that the problem should reflect social approach or some needy solution.
- Also, ensure that the problem focuses existing problem in the research domain/subject
- Include time frame over which the problem has been occurring
- Quantify the problem
- One should be able to ask the question related to the problem (why and how?)
- Problem statement should be defined as you start reading the literature root level

2.8.1.1 Review the Problem Statements Based On

- It should focus on only one problem
- It should not be more than two statements
- The problem statement should not suggest the solution

2.8.1.2 Poor Way of Writing Problem Statements

The problem statement "*Novel algorithm to improve the speed of classification using K-means clustering method*," is an inappropriate one. In this, the "K-mean" clustering method is given in the problem statement itself, which is wrong. When one writes the specific method then he/she has to use that method only for experimentation. It limits the scope of the research. As said earlier, research is not time bound. While pursuing research, the researcher goes on doing experimentation, he/she may devise new methods or techniques for getting better result. So, one should not specify the method in the problem statement itself. Also in this statement "novel" and "algorithm" words are used, which are to be avoided. As the research is always used to devise novel/innovative/new method/algorithm, for improving or optimizing some parameters. Therefore one should not write the words such as "novel" and "algorithm" in the problem statement itself.

Another example of poor problem statement is *"Finding/identifying the defects in the fetus using video enhancement."* In this statement, "fetus" is used, but the video or images of a fetus are not allowed legally. So while determining the problem statement, words such as "fetus" should not be used, as the database for fetus is not permitted legally.

2.8.2 Errors in Defining the Problem Statement

If the sufficient time is not given by the researcher for deciding the problem statement, then there may be errors in defining the problem statement. The researcher has to take utmost care and provide sufficient time in defining the problem statement to pursue. When researcher fails to pursue the following important points, then there are errors in defining the problem statement. These errors are generally due to the following points:

- Researchers don't have an idea about what exactly to do (pinpointing of the research subject)
- Researchers are not qualified in the subject to carry out research design
- Why you feel that the topic is innovative or meaningful or of interest to the research
- Availability of database for the selected research subject
- Any legal issues related to the research domain
- Any issues related to the safety (such as military applications, etc.)
- No consideration of the cost involved while pursuing the research
- Too narrow or too vague problem statement
- Don't select a controversial problem
- Don't choose a subject/topic that is overdone; it is difficult to throw new light on such cases

If these points are not considered properly and the researcher has not given any thought related to them, then there are chances of errors in defining the research problem.

2.9 Summary

Literature survey and its review plays an important role in the definition of the problem statement. As a result, the different ways of performing the literature survey and review is explained in detail. This chapter is a treasure for those who would like to understand the concepts behind the literature survey and review. The different ways of collecting the information or data through interviews, focus groups and questionnaire are explained. Based on literature survey and review the researchers should be able to define their own problem statement. Hence this chapter can serve as an important chapter for identifying problem statement for post graduate students as well as research scholars.

Further Reading

1. Abdulla Ramdhani, Muhammad ali Ramdhani, Abdusy Syakur Amin, "Writing a Literature Review Research Paper: A step by step approach" International Journal of basic and Applied Science, Vol 03, No 01, pp.47–56, July 2014.
2. Justus J. Randolph Walden University, "A Guide to Writing the Dissertation Literature Review, A peer-reviewed electronic journal, Practical Assessment, Research & Evaluation", ISSN 1531–7714, Volume 14, Number 13, pp 1–13, June 2009, ISSN 1531–7714.
3. D. R. Rowland, The Learning Hub, Student Services, The University of Queensland, Reviewing the Literature: "A Short Guide for Research Students", pp1–20.
4. Y. Chen, L. Zhang, B. Lin, Y. Xu, X. Ren. Fighting Detection Based on Optical Flow Context Histogram. Proc. of IEEE 2nd Int. Conf. on Innovations in Bio-inspired Computing and Applications, pp. 95–98, 2011.
5. Rub´en Heras Evangelio and Thomas Sikora, "Complementary Background Models for the Detection of Static and Moving Objects in Crowded Environments", 8th IEEE International Conference on Advanced Video and Signal-Based Surveillance, 2011
6. Pierre-Marc Jodoin, Venkatesh Saligrama, Janusz Konrad, "Behavior Subtraction", IEEE Trans. Image Process, vol. 21, no. 9, pp. 4244–4255, September 2012.
7. Dawei Li, Lihong Xu, Erik D. Good Man, "Illumination-Robust Foreground Detection in a Video Surveillance System", IEEE Trans. On Circuits and Systems for Video Technology, vol. 23, no. 10, pp. 1637–1650, October 2013.
8. Kalyan Kumar Hati, Pankaj Kumar Sa, Banshidhar Majhi, "Intensity Range Based Background Subtraction for Effective Object Detection", IEEE Signal Processing Letters, vol. 20, no. 8, pp. 759–762, August 2013.
9. R. Cutler, L. Davis, "Robust Real-Time Periodic Motion Detection, Analysis, and Applications", IEEE Trans. Pattern Anal Mach. Intell, vol. 22, no. 8, pp. 781–796, August 2000.
10. L. Zhang, S. Li, X. Yuan, S. Xiang, "Real-Time Object Classification in Video Surveillance Based on Appearance Learning", Proc. IEEE Int. Workshop Visual Surveillance in Conjunction with CVPR, 2007.
11. M. Kafai, B. Bhanu, "Dynamic Bayesian Networks for Vehicle Classification in Video", IEEE Trans. Ind. Inf., vol. 8, no. 1, pp. 100–109, February 2012.
12. Tianzhu Zhang, Si Liu, Changsheng Xu,Hanqing Lu, "Mining Semantic Context Information for Intelligent Video Surveillance of Traffic Scenes", IEEE Transactions on Industrial Informatics, vol. 9, no. 1, pp. 149–160, February 2013.
13. Q. Chen, Q.S. Sun, P.A. Heng, D.S. Xia, "Two-Stage Object Tracking Method Based on Kernel and Active Contour", IEEE Trans.Circuits Syst. Video Technol., vol. 20, no. 4, pp. 605–609, 2010.
14. Y. Chai, S. Shin, K. Chang, T. Kim, "Real-Time User Interface using Particle Filter with Integral Histogram", IEEE Trans. Consum. Electron, vol. 56, no. 2, pp. 510–515, 2010.
15. J. Chiverton, X. Xie, M. Mirmehdi, "Automatic Bootstrapping and Tracking of Object Contours", IEEE Trans. Image Process, vol. 21, no. 3, pp. 1231–1245, 2012.
16. S. Vishwakarma, A. Sapre, A. Agrawal, "Action Recognition Using Cuboids of Interest Points", IEEE Int. Conf. on Signal Processing, Communications and Computing (ICSPCC), Electronic ISBN: 978-1-4577-0894-7, pp. 1–6, 2011.
17. Mahfuzul Haque, Manzur Murshed, "Robust Background Subtraction Based on Perceptual Mixture-of-Gaussians with Dynamic Adaptation Speed", Proc of the 2012 IEEE International Conference on Multimedia and Expo Workshops (ICMEW '12), Melbourne, Australia, pp. 396–401, July 2012.
18. Ronald Poppe, "A Survey on Vision-Based Human Action Recognition", Image Vision Comput. J., vol. 28, no. 6, pp. 976–990, June 2010.
19. Saad Ali, Mubarak Shah, "Human Action Recognition in Videos Using Kinematic Features and Multiple Instance Learning", The IEEE Transactions on Pattern Analysis and Machine Intelligence, vol. 32, no. 2, pp. 288–303, February 2010.

20. Carolina Garate, Sofia Zaidenberg, Julien Badie, Francois Bremond, "Group Tracking and Behavior Recognition in Long Video Surveillance Sequences", VISAPP – 9th International Joint Conference on Computer Vision, Imaging and Computer Graphics, January 2014.
21. Hala H. Ahmed Taha Zayed, et al., "On Behavior Analysis in Video Surveillance", Proc of The 6th International Conference on Information Technology, 2013
22. Sofia Zaidenberg, Bernard Boulay, et al., "A Generic Framework for Video Understanding Applied to Group Behavior Recognition", International Conference on Advanced Video Signal Based Surveillance (AVSS), 2012.
23. Nick C. Tang, et.al, "Robust Action Recognition via Borrowing Information Across Video Modalities", IEEE Transactions on Image Processing, vol. 24, no. 2, pp. 602–614, February 2015.
24. Tian Wang, Hichem Snoussi, "Detection of Abnormal Visual Events via Global Optical Flow Orientation Histogram", IEEE Trans. On Information Forensics And Security, Vol 9, No 6, June 2014.
25. Jasper R. Van Huis, Henri Bouma et.al., "Track-based event recognition in a realistic crowded environment", In proc of Society of Photo-Optical Instrumentation Engineers, SPIE, 2014
26. Maria Anderson, FredrikGustafsson et. Al, "Recognition of Anomalous Motion patterns in Urban Surveillance", Selected Topics in Signal Processing, IEEE Journal Volume:7, Issue: 1, 22 January 2013.
27. Xu, M., Zuo, L., Iyengar, S., Goldfain, A., DelloStritto, J, " A semi-supervised hidden Markov model-based activity monitoring system.", In 33rd Annual Int. Conf. of the IEEE Engineering in Medicine and Biology Society (EMBC), Boston, Massachusetts USA, pp. 1794–1797 (2011).
28. Vishwakarma, S., Agrawal, A.: A novel approach for feature quantization using one-dimensional histogram. In: Annual IEEE India Conference (INDICON), pp. 1–4 (2011)
29. Yan Song et.al, "Describing Trajectory of Surface Patch for human Action Recognition on RGB and Depth Videos", IEEE Signal Processing Letter, Vol. 22, No. 4, April 2015.

3

Research Design

Prachi Joshi

CONTENTS

- 3.1 Introduction ...70
  - 3.1.1 What Is Research Design? ...70
  - 3.1.2 Necessity of Research Design ...71
  - 3.1.3 Framework of Research Design ...71
  - 3.1.4 Parameters of a Research Design ...71
  - 3.1.5 Design and Methods ...74
- 3.2 Approaches of Research Design ...75
  - 3.2.1 Qualitative versus Quantitative Research ...75
- 3.3 Types of Research Designs ...75
  - 3.3.1 Explanatory Research Design ...75
  - 3.3.2 Descriptive Research Design ...77
  - 3.3.3 Diagnostic Research Design ...77
  - 3.3.4 Experimental Research Design ...77
  - 3.3.5 Exploratory Research Design ...78
  - 3.3.6 Hypothesis-Testing Research Design ...78
- 3.4 Principles of Experimental Design ...79
  - 3.4.1 Principle of Replication ...79
  - 3.4.2 Principle of Randomization ...80
  - 3.4.3 Principle of Local Control ...80
- 3.5 Design of Experiments ...80
  - 3.5.1 Post-Test Only Design ...81
  - 3.5.2 Pre-Test Post-Test Only Design ...81
  - 3.5.3 Completely Randomized Design ...83
  - 3.5.4 Randomized Block Design ...84
  - 3.5.5 Latin Square Design ...85
  - 3.5.6 Factorial Design ...86
  - 3.5.7 Quazi-Experimental Design ...87
  - 3.5.8 Cross over Design ...88
- 3.6 Sampling Concepts ...88
  - 3.6.1 Design of Sample ...89
  - 3.6.2 Principles of Sampling ...89
  - 3.6.3 Preparing a Sample Design ...90
  - 3.6.4 Sampling and Nonsampling Errors ...91
  - 3.6.5 Selection of Sampling Procedure ...91
- 3.7 Methods of Sampling ...92
  - 3.7.1 Probability Sampling ...92

3.7.2 Nonprobability Sampling. 96
3.7.3 Sampling in machine learning approaches. 97
3.8 Summary 98
Further Reading. 98

> The Twelve Principles of Character Are Honesty, Understanding, Compassion, Appreciation, Patience, Discipline, Fortitude, Perseverance, Humour, Humility, Generosity and Respect.
> —Kathryn B. Johnson

Learning objectives of this chapter are to:

- Understand the importance of research design
- Analyse the various types of research designs and select appropriate design for experimentation
- Distinguish and learn the applicability of the experimental designs
- Investigate the sampling techniques and select suitable method for research

This chapter will enable the researcher to:

- Use the research design in the proposed research experimentation
- Select and apply the design as per the need of the proposed research
- Identify and use the sampling method as per the requirement
- Exploit the research design for improved research findings

3.1 Introduction

Any research implementation needs a systematic approach toward development. This is important from the view of exploring right directions to come up with concrete results. So what relationship exists between the methodology and the design? Which aspects of design are required for a problem at hand? What about the samples used? How to select them? Let us understand. The chapter is split into two major sections, research design and the sampling concepts.

3.1.1 What Is Research Design?

At the heart of the research lies in the understanding of the problem. A researcher is capable to capture and identifying the need of the research and formulate the problem definition. This is the most preliminary step but lot of questions that need to be addressed are -How to carry out the research? How to practically deploy solutions and deliver accurate results?

A research design is the most important step in giving a direction to the research problem. It is the overall plan that deals with the aspects of complete design from the study type, data collection approaches, experimental designs, and statistical approaches for data samples.

3.1.2 Necessity of Research Design

Why is the research design so very important? A researcher can logically decide and select on the methods, build his/her own plan and work. But it does not happen this way. For a researcher of any field, the design guides him from the perspective of the type of data that can be used, the collection methods which can be looked at, to determine methods suitable for the problem at hand and finally able to have concrete deliverables!

In daily life too, there exists a proper planning; a to do list possibly that every one of us makes and follows some plan to get a "tick mark" on those lists. Similarly, a research design guides us to overcome the difficulties that one is likely to face during the research to get outcomes. Whether one is working in mechanical industry or it could be on the IT product development, a mind-map needs to be prepared. The dependencies of the components, the requirements of experimentation and research, the data/equipment availability, the economic front and other things are required to be concentrated upon for coming up with effective results. A failure in planning of any of the aspects could land up causing an irrevocable error.

The research design helps the researcher to understand the dependencies, consider the overall road map for carrying out the research along with identifying the minute details. So, in a nutshell, it is extremely necessary that a research design should be prepared meticulously and exploited for the betterment of a benchmark research.

3.1.3 Framework of Research Design

From the preceding points, one thing is clear that the design actually facilitates the research toward a right direction. So essentially a research framework deals with:

(i) Defining precise, up to the point problem statement
(ii) Approaches/techniques that can be put to practice to collect samples/data
(iii) The details about the data that need to be analyzed and further researched upon
(iv) Approaches/experimental setups to be executed for the data processing and analysis.

One thing that a researcher need to keep in mind is that all types of research designs (which we would be studying further) are not applicable to all types of research problems and vice-versa is also not true. The type of research study would be the one determining the design for it.

In general, without belonging to any specific category of research, the framework is constituted of components as shown in Figure 3.1.

3.1.4 Parameters of a Research Design

In this section, we will discuss some important terminology used while developing a research design. Based on the lines of framework, how the entire research would be placed is detailed as below:

(i) Title of research:

Every researcher wants to focus, develop, and deliver a very potential research component. He is keen to showcase the necessity of the proposed work. This needs

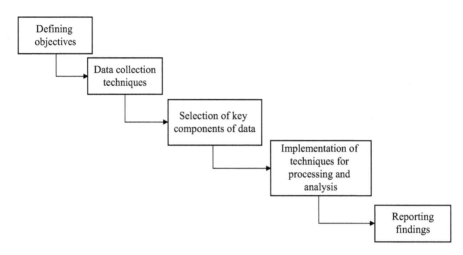

FIGURE 3.1
Research Design Framework

to be made clear by properly defining the title of the research. There should be no ambiguity; needs to be concise and clearly state what the researcher plans to do.

(ii) Importance of the research:

The title gives a first-level idea about the research but the importance of the research in a more elaborate way should be mentioned and the researcher himself should be well aware of the necessity. This section puts forth the significance of the work being carried out.

(iii) Literature survey:

While coming up with research design it is equally necessary to understand, relate and study what other researchers have contributed to. So, an understanding of which reference material needs to be applied and studied further on along with identification of shortfalls other researchers is important from the researcher's perspective. This aids in defining the scope of research and to have achievable objectives.

(iv) Scope/objectives and problem statement:

The design should put forth the clarity on the objectives and a concrete problem statement. A clear idea about the tasks that are required to be carried out as a part of research should be explicitly mentioned with the objectives. A problem statement defines the outcomes expected out of the research in a very precise way. The statement should be written in concise way and as such should be able to put light on the issues that the researcher will be addressing. Literature survey shows way to define the problem statement.

(v) Concepts of the terminology used:

A key aspect while coming up with research design is that the researcher needs to have understanding of the terms he would be using in the research. In this sub-point, we will high light the terms with regard to the research design:

- Variables – An item used in research that can take up different values. These values can be quantitative like temperature (Celsius/Fahrenheit), measurement of earthquake (Richter scale) and so on. Variables could be continuous – numerical values as mentioned earlier or categorical-qualitative. Qualitative variable can be quality of air (good/bad). They are further classified as dependent and independent. A relation between weight and height of an individual – these variables can be termed as dependent variables. Relation between amount of rainfall and height of an individual – independent variables
- Extraneous variable – independent variable that may not affect the research directly but could be a governing factor for a dependent variable is termed as extraneous variable. For example, a relation between time spent on social media and the type of person (introvert or extrovert). Time spent on social media and type of person are two independent variables. But time spent on social media is dependent variable for an extraneous variable – time spent on exercising. So, a third variable affecting this which is actually not considered in the research study. One needs to be very careful thus while framing the statements and understanding the extraneous variables
- Control – term typically used in research design to reduce the influence of extraneous variable
- Confounded relationships – If the dependent variable is restricted by the extraneous variable, it is confounded relationship
- Hypothesis – one of the most important aspect of research design. Hypothesis is a probable or a tentative solution to a given problem. It is a claim that a researcher needs to test prior to coming to a conclusion that it is valid. No researcher can proceed without stating a hypothesis
- Experimental and control groups – While dealing with experimental designs, when a group of samples is experimented with normal or traditional conditions, it is referred to as control group whereas when a group is experimented with new/certain specific conditions/environment it is termed as experimental group. These would be dealt in detail in the experimentation design section later
- Treatments – The varied conditions to which the above-mentioned groups are exposed to are referred to as treatments

(vi) Data – Selection and collection and analysis:

While one is ready with the concepts and the research topic, as mentioned in the research design framework, one has to narrow down on the data selection and collection methods. The amount of data viz. number of samples, the type of data, the testing data to be used should be decided prior to the commencement of research. This is dependent on the nature of research like if it's a research based on description – it would need large amount of data as compared to others. Data collection is the most time-consuming task. Many times the researcher faces a problem of unavailability of the data. In this design phase, following factor should be taken care of.

- from where to get the data – reaching out to experts and getting complete understating of the data to be used during the research process

- how to collect – whether in the form of questionnaire or surveys or any other method
- whether to use any specific equipment/new online forms to gather the required data
- type of collection process – whether it is required throughout the research or the data needs to be acquired only in the beginning. This is again characterised by the nature of research one would be doing.

One of the biggest challenges that a researcher often faces is getting the accurate and precise data. Hence data acquisition is a very essential step that needs to be addressed during the design stage.

Although analysis is dealt with while the research is ongoing, the planning about what analysis needs to be performed should be done prior. This is to get an overall idea about how the research would proceed and what needs to be done.

(vii) Results: Interpretations and conclusions

Why are we discussing about results at this stage? A very crucial aspect that a researcher should know is; how he plans to deliver the results and state his/her claims while interpreting the results. Though a variety of methods – graphs, tables, scatter plots etc. can be used – the way the results are populated, what factors need to be concentrated upon while talking about results, any specific parameters that can been further taken up to gain more insights need to made clear. With reference to the previous base cases and earlier research, the new claims should be justified to derive conclusions. This planning about the parameters that one expects at the result should be done in the beginning – and hence in the research design. A detail justification of the expected results and the actual results needs also to be added at the end and researcher cannot leave it without giving an answer to "why" the difference was identified.

(viii) Further research:

Once we begin with the planning of objectives to data collection, the analysis to results, the other aspect that one can specify is the further research. When one is defining the scope, a researcher gets the clarity of thoughts as to which particular arears he would be exploring and what can be done ahead which he is not accounting into his/her scope. Of course, many times despite mentioning about the further research or areas out of scope in this design stage, the picture gets clearer during or when the research is coming to an end. Despite this fact, it is a good practice to mention about it.

3.1.5 Design and Methods

One fundamental aspect one needs to take care of is to understand the difference between design and method. Design is not any method that one needs to carry out. It is a way of planning things. No explicit method in research design talks about the way in which data collection should occur or any specific approach to be applied. The fact is that it is a plan that deals with the different research areas that would be experimental or analytical. There are different approaches or methods that are performed or used for different design types. For example, a design type can be "case study." For such a design, one could go with detailed data collection with online forms or questionnaire and then go ahead with any method the researcher finds appropriate to get the analysis.

3.2 Approaches of Research Design

What is it that one wants to come up with? Essentially what philosophy exists in approaches while carrying out the research? Let us explore more.

3.2.1 Qualitative versus Quantitative Research

A measurable amount is referred as "quantity." So, something that one can express in the results with a specific amount that could be compared can be called as quantitative research. Imagine a researcher trying to gain insights about the amount of pollution and its percentage impact. Such a research is a case of quantitative research- that has rigorous experimentation and a lot of findings. So, for the said example findings can be different plots with percentile impacts on people in different geographical areas and so on.

In case of qualitative research, its deals with subjective analysis. More of psychological aspects involved here. A survey about how much people are depressed if they spend time on social media can be an example of qualitative analysis. We can relate it to the behaviour or cognitive sciences. A very common example we can say about qualitative analysis is of social network sites. We often get recommendations for some specific posts. Though there is some sort of intelligence embedded in the methodology that is used, but the outcomes are qualitative. We get to know the kind of person, his/her likes, dislikes and so on.

Yet there has been a long debate as to whether research design is actually getting impacted with the qualitative or quantitative research. With the reference to case studies, the researchers claim that it is actually insignificant to classify the design into qualitative or quantitative.

3.3 Types of Research Designs

This section deals with the different research designs. Addressing the different ways in which the data collection affects and how the type of research is formulated and takes place contribute in formulation of a research design. One point that every researcher has to understand is that the design framework described earlier is a baseline. Depending on the type of research, the design has to be moulded to have adoptions of different variants. This could be for hypothesis building, the data collection or any other step.

3.3.1 Explanatory Research Design

As the name suggests, this type of research design addresses the "Why" questions. Consider a research area where one has to study as to why majority of students are keen in pursuing engineering degree despite several other avenues. This question changes the way in which data collection occurs. Detailed data collection for the problem at hand needs to be carried out while creating the research design.

While carrying out explanatory research, we need to understand causal relationships. Since it's all about addressing the question of why we need to determine the impacts and effects of one attribute with other that is involved in the research.

While saying so, the causal relationships are typically given as X → Y, where Y is some attribute or characteristic that is affected by X. Causal relationships can be direct, indirect, or complex, or a combination of direct and indirect.

For example: in case of direct relationship, educational qualifications will have impact on job opportunity. Whereas in case of indirect relationship, it can be represented as a causal chain such as many in-between links from educational qualifications to domain areas and then job opportunities.

While talking about complex relationships, one link can be established from Educational qualification to internships and finally to job opportunity where as other link can be from Educational qualification to domain area and then job opportunity. Figure 3.2 depicts the relationships.

While dealing with explanatory relationships, it is extremely necessary to carefully understand and identify the relationships. Many times, a researcher can land up in selection of parameters which are insignificant but due to some changes that have occurred in a particular parameter, they then need to consider it to be a part of causal relationship. Since we are trying to have a reasoning justified in this kind of design, the factors which are influential and are not related; if are yet counted in the experimentation, they may lead to wrong findings.

For example, in recent years the number of people who have started living in old age homes has increased at the same time number of youth that has gone abroad for work has increased. It is not necessary that the youth taking up jobs abroad has caused the

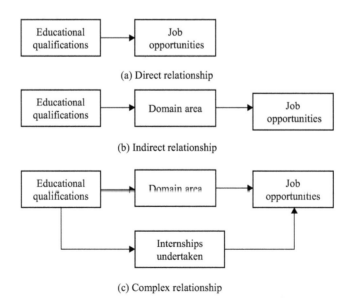

FIGURE 3.2
Causal Relationships

count of people in old age home to increase. There could be other factors too. So, data parameters for reasoning are very much important!

While it is equally important that while working on explanatory relationships, wrong inferences or conclusions to the "why" questions should be omitted in the design.

3.3.2 Descriptive Research Design

When we talk about descriptive research design, it deals with finding out the characteristics or particular behaviour or pattern in a specific group. It could be a case where a researcher is trying to find out some outlier by performing the study of the group of data points.

This study deals with coming up with findings in predictive way to understand the different/unusual characteristics or unnoticed behaviour. The research enlists, describes the reasons and the observations.

While preparing design for this type of research, one has to definitely follow the formal earlier explained process but the basic research design needs to ensure that a comparative study is also added up. Since the study deals with observing the peculiarities and publishing their findings, the data collection here can include previously recorded data, questionnaire, or any other means of getting authentic data.

3.3.3 Diagnostic Research Design

Diagnostic research deals with finding out associations. More specific, it is trying to conclude with findings that establish relationships. We can say about it as one aspect that exists and has led to other aspect. A typical example for association can be the findings of research in a supermarket shopping. It can be study of establishing associations about people who tend to buy butter along with bread or even associations of buying eggs along with milk. Such a study involves mining associations.

A research design for this analysis should be carefully worked on. It is necessary to get data samples that suit or are belonging to the category of association the researcher is trying to explore. So, again in design where we are following the typical research design framework, the design needs to accommodate and perform changes as per the type of research. So, in this diagnostic category, the sample design needs to be prepared to accurately get the data set required for study. Many researchers would be keen on using the different sampling techniques – random, cross-fold, or any other. A good mixture of samples are necessary while doing this type of research.

While describing the diagnostic research, there are a lot of similarities between the descriptive one and diagnostic type. Even the sample design to populate the data, the comparative analysis to understand associations, the tabulation of results and their formats can be very much similar and one could see a significant overlap between these two designs.

The major difference in these two studies is the methods that would be worked on and finally the conclusions are different.

3.3.4 Experimental Research Design

Whenever we perform any type of experiment, we tend to infer or come up with something novel. As a researcher, knowing the involved parameters and their limitations, we make attempts to determine some new things. When the research is of such a category, the research design has to ensure that, it makes use of following parameters in the planning stage. It involves validations and full proof conclusions to portray that the experiment is valid. We can relate this type to hypothesis testing research as well.

In this type of research design following things are accounted:

(i) Problem formulation
(ii) Defining the hypothesis
(iii) Preparation of the experimental design – this includes the variables, the quantities, their relations, their properties, the environment in which the experiment is to be performed, proper selection of the equipment for experimentation and selection of any other material/substances in the procedure.
(iv) Analysis of the impact and effect on the variables post experimentation.
(v) Validation of the experiments with statistical reasoning and testing.

There are different variants and principles in case of experimental design. They will be discussed in a later section in this chapter.

3.3.5 Exploratory Research Design

Let us discuss this type of research design by an example. Assume you are fond of eating. There is a typical hotel, one of your favorites, and you often visit that hotel. This is where you are exploiting a known place. Now if there is a new hotel that has opened and if you go to this new hotel rather than your regular one – you are exploring! Maybe you would like or you won't like the food there but you never know unless you go there!

So, in exploratory design, the bonus is on inventing or discovering new things!

In such a research, the design needs to be very flexible. The most important factor in exploratory study is the survey!

While doing exploratory study, surveys are very much important to lead to a strong hypothesis. Surveys can be in the form of literature review so that the previous findings assist the researcher in concise definition of the problem statement. Many times online surveys or questionnaire where researcher gets data filled from experts helps in analysis of the real issues. This type of survey is named as experience survey. There could be other approaches like – selection of data samples that carry potential information for further exploratory analysis. This can be merged with the ones discussed earlier. Many times interviews too are used as a collection approach for obtaining critical information required in the exploratory analysis.

> So, to summarize – exploratory is all about – findings with new tasks associated with the dependencies along with lot of surveys!!!

3.3.6 Hypothesis-Testing Research Design

In case of hypothesis testing, the study deals with finding out the causal relationships between the variables associated. Generally, a researcher is trying to explain or claim dependency of "Y" (a dependent variable) on "X" (an independent variable). Such a relation can be put forth in the form of hypothesis.

$F(X) = Y$; where X: independent variable and Y: dependent variable

Research Design

So, while defining the hypothesis, it needs to be very precise and clear with no ambiguity. As an example, the hypothesis can be "Students pursuing Masters in Engineering are likely to buy this book of Research Methodology than students studying other streams," Such statements need to be tested against the evidence, need performing experiments to validate the claims to infer about the dependencies. So, often while talking about hypothesis testing, it's about the experimentation design.

This design was evolved owing to the work by Professor R. A. Fisher, who carried out experimentation on the agricultural land and later came up with the notion for hypothesis testing. He divided the land and set up experiments and inferred on yield and other parameter and variables. It all motivated him to formulate the hypothesis to determine this outcome. Hence, hypothesis testing makes use of experimental designs!

Figure 3.3 depicts key aspects of the different research designs.

3.4 Principles of Experimental Design

Though a brief introduction to experimental designs has been done in the previous section, let us understand in detail what are the principles to deal with such a design. While developing a research design, a lot of experimentation needs to be carried out. But while doing so, there are basic principles that need to be followed to validate the results of the experimentation. Let us understand them.

3.4.1 Principle of Replication:

To put forth results on any experimentation, performing the experiment just once is not sufficient. The principle of replication states this – to perform the experiment number of times and not relying on the output obtained after performing it just once.

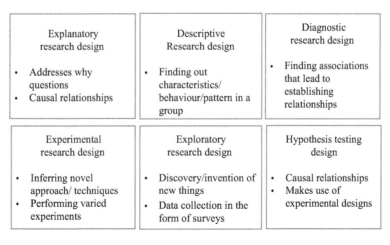

FIGURE 3.3
Summary of the Research Designs

For example, we want to test the impact of eating two types of fruits – apples and bananas on kindergarten school children. One can think of different ways to carry out this experiment. Assume that we split the kids and form two groups and test them by giving one group bananas and other group apples. This is just one experiment that we performed. But instead if we split them into multiple sub-groups and compare the results – we are trying to replicate it! It's not sufficient, the same experiment can be carried out multiple times may be for few days or months. What we get at the end of experimentation by replication is more precision. Accuracy for drawing conclusion to justify the claims! Do remember replication is computationally expensive!

3.4.2 Principle of Randomization:

In the replication principle we discussed about repeating the same experiment – simultaneously with the groups formed and also repeating it for a period of time, but what matters also are the samples in the group. While forming the groups – during the Replication principle, if same groups are used for the experimentation n number of times, the results could be biased. What is essential is that we have random samples in the groups. So, the sub-groups will have different kids for an experiment to replicate rather than going with fixed subgroups. By doing this in each turn of experiment, the results can be statistically justified and moreover are generalized over the randomized samples.

3.4.3 Principle of Local Control:

Replication talks about performing repetitions of the same experiment, randomization talks about selection of random samples, and local control talks about reducing the experimental error.

Typically, while carrying out the experimentation, there are factors known as extraneous variables which would cause high error rate. In our example – obese kids can be thought of as an extraneous variable. So even though while forming groups with random samples, the obese kinds falling in one subgroup could impact the experimentation outcome. Hence varying this variable more; can help in reducing the error. So, initially with homogeneous groups been formed, which are referred to as "blocks" – they are further divided by taking into account this factor of obese kids. Then the experimentation variables are assigned to measure the outcomes.

3.5 Design of Experiments

This section discusses the different variants in which experimentation can be structured. Typically, in experimentation designs, the samples are split into the experimentation or test group and control groups (refer to Section 3.1.4). The outcomes of the experimentation are measured based on these groups.

3.5.1 Post-Test Only Design

Any experimentation design often has the first part as splitting of the samples – samples to be assigned to the experimentation group and control group. After this the test is carried out. In post-test, the treatment (as discussed in earlier section, the conditions to which the groups are exposed is treatment) is then introduced to the experimentation group only. The impact of the treatment is obtained by calculating the difference between the dependent variable from experimentation group and control group.

One major consideration under this design is it is assumed that the group's performance toward the treatment is similar. Unlikely that the assumption would fail, an extraneous variable could be encountered in the treatment.

For example, to study the impact of introduction of a soft skills workshop for second-year students. In the Experimentation group, the students are exposed to the treatment, that is, workshop is conducted along with the regular lectures whereas the control group just follows the regular lectures. An evaluation for the outcome measurement post treatment is done. So the performance of the control group at the end and the performance of the experimentation group at the end is evaluated. The difference between these outcomes will express the impact of conduction of soft skills workshop.

Figure 3.4 shows the post-test control design. As we can observe the outcome is the measure of X–Y, where X is control group dependent variable and Y is the experimentation group. Many times conducting a pre-test is expensive and hence only a post-test is conducted.

3.5.2 Pre-Test Post-Test Only Design

There are two variants of this design:

(i) without control group
(ii) with control group

In case of without control group, following steps are applied:

- The experimentation group is selected and is only taken for treatment

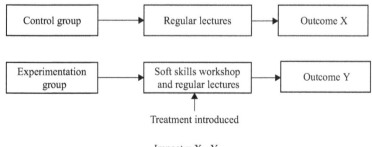

FIGURE 3.4
Post-Test Control Design

- The outcome of the treatment is measured on the basis of the difference between the dependent variable of only this group, that is, prior to treatment and post treatment

A major drawback of using this is with time delays, extraneous variables could impact!

In the second case, both the groups are used in the design. Here, the steps are:

- The value of the dependent variable is calculated prior to the treatment and then post treatment
- Only the experimentation group is exposed to the treatment
- The difference between the changes in the dependent variables of the experimentation group and the control group are noted to determine the outcome impact

Example for case 1:
Let us take the same example discussed earlier. In step 1, the pre-test measure is taken of the experimentation group, that is, what was the student's performance prior to the soft skills workshop. In step 2, post-test measure is taken, that is, performance after the workshop. The difference between these two outcomes (performances) will establish the impact of soft skills workshop conduction. This is the case of without control group.

Example for case 2:
It's a different way- the pre-measure is taken of both the groups i.e. control and experimentation.

Step 1. The student's performance prior to exposure to the treatment is calculated for control as well as experimentation group.

Step 2. Then only the experimentation group is given treatment (that is workshop), whereas the control group goes ahead with its regular lectures.

Step 3. After the treatment, the performance is measured for both groups.

Step 4. To determine the final impact, initially calculated difference between the outcome of control group (pre and post) is taken and difference between outcome of experimentation (pre and post) is calculated.

Step 5. After this, the final difference between both outcomes determines the impact of the treatment.

Figure 3.5a and b shows the example with and without control group.

(a)
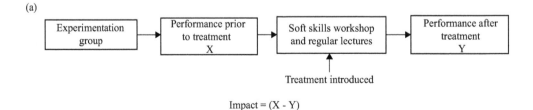

Impact = (X - Y)

Research Design

(b)

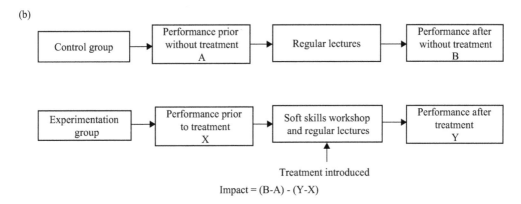

FIGURE 3.5
(a) Without Control Group. (b) With Control Group

3.5.3 Completely Randomized Design

After having understood the pre and post-test designs, there are a few aspects in designing an experiment which are left out.

In this type of design, there are two categories or types in which the sample assignment to both the groups occurs:

- Two group randomized design
- Random replications design

(i) Two group randomized design

Two group randomized design one of the simplest forms of design. It states that there should be random assignment of the selected samples to the two groups and random selection of the samples from the entire sample set/population. So, for example, if we are to study impact of introduction of soft skills workshop of second-year students, from the entire population, we can select any 20 – randomly (first level random selection) then assign any 10 randomly to control group and 10 to experimentation group (second level random selection).

In this type of design, the treatment can be applied to both the groups. Most often preferred and used by researchers in the behaviour sciences. It can be used along with the pre-test post-test only. Figure 3.6 depicts this design.

One of the major concern factor here is the impact of the extraneous variable leading to confounding relationships.

(ii) Random replications design

There could be many drawbacks for the two group randomized design. As mentioned, the extraneous variable and moreover the way in which the samples are split. Just random selection of the samples and then applying the treatment to both the groups will lead to biased outcomes. It is most sensible to have repetitions of the two group randomized design. In random replications design, the sample set is considered to be available as study group and experiment group; from this further random selection for "selected group" – study and treatment takes place. Then it is assigned

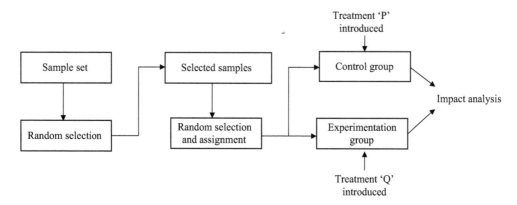

FIGURE 3.6
Two Group Randomized Design

to the Experimental and Control group – that too randomly!! The samples are assigned to "n" number of Experimental and Control groups ("n" will be dependent on the nature of experimentation). So, if "n" is three then, three experimental groups and three control groups will be formed. This leads to the creation of groups with reduced bias.

So, adding replications to the previous design, here for the same example where we have to study impact of introduction of soft skills work shop of second year students, from the entire population, we will first split it into two parts. It is possible that we already have two different populations. So, assuming that we have 50 samples in each group – study and experiment – initial population.

Post this, random samples will be selected – forming "selected group". So, from 50, maybe 25 samples are selected in the study and experiment group.

Further from these "selected groups," again random samples will be selected to finally assign to experimentation and control group. So, may be samples are selected randomly to be now assigned to the experimentation and control from the previous step. Remember there are "n" groups formed with "n" samples assigned for the groups.

Then each of this experimentation and control group now undergoes the treatment.

Figure 3.7a and b depict the working of the random replications.

3.5.4 Randomized Block Design

In case of randomized block design, this design is used when there are more number of treatments. The samples are clubbed to form blocks. Each block has x number of samples. The formation of the blocks is such that each block would comprise of homogeneous samples. That means the bias within the block is made minimal. Now, for every block, there are set of treatments that have to be taken. For every block, if randomly the treatment is undertaken by the samples, then it is called as randomized block design (R.B.) design.

For example, if the number of treatments is five and the number of blocks is four, Figure 3.8 shows possible assignment and the way in which the treatments can be taken up.

Research Design

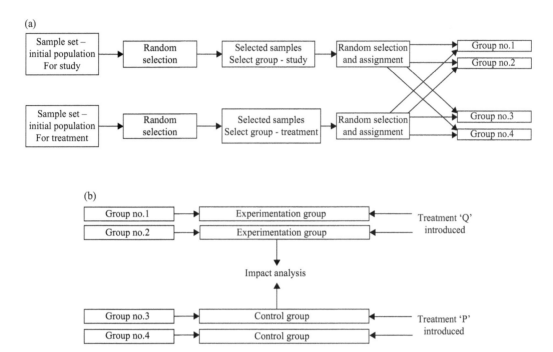

FIGURE 3.7
(a) Working of the Random Replications. (b) Working of the Random Replications

3.5.5 Latin Square Design

Typically, in a R.B. design, the treatments are assigned randomly to the blocks. Here every treatment gets assigned once to a block, unlike in completely randomised design. To overcome the cost factor involved in the replication and assignments involved in R.B. design, Latin Square design is used. This type of design is best suited in agricultural research. The central theme is to split the experiment data into rows and columns such that each row and column will have same number of treatments being assigned.

So, Latin Square with order "x" will bear x symbols such that it equals to x^2 cells. Each cell will be of x rows and columns and will carry the symbol only once.

A Latin Square of order 4 is given here:

LS1	LS2	LS3	LS4
LS2	LS3	LS4	LS1
LS3	LS4	LS1	LS2
LS4	LS1	LS2	LS3

Consider an example where the impact of use of mobile phones on students on different days is to be studied. So, we have different mobile phones, differences in the way students use it and differences in which their days are tied up.

	TREATMENTS					
Block 1	T3	T1	T4	T2	T6	T5
Block 2	T4	T2	T6	T5	T3	T1
Block 3	T2	T3	T1	T4	T5	T6
Block 4	T4	T2	T5	T6	T1	T3
Block 5	T6	T5	T2	T1	T3	T4

FIGURE 3.8
Sample Example for Randomized Block design

Assume it is to be done for five days – Monday to Friday. Similarly, let us say we have five students and five mobile phones. In such a scenario for a complete replication design, one would need 5 * 5 * 5 experiments! Let us begin with 25 experiments. This is done with Latin square as follows in Figure 3.9.

As shown in this figure, consider the first entry. Here the student 1 will be using P mobile on Monday. Similarly, for other cells, the assignment is done. If we happen to select a different Latin square, we would be getting different outcomes, that is, the combinations would differ.

To summarize, the treatments are replicated with rows and columns. The analysis in this design is dependent on the selection of the square and the impact/conclusions tend to vary with different selection. Moreover, an assumption is taken here that there is no interaction between the cells. Analysis from Lantin Square is comparatively simple. But it does need the treatments to be equal to the number of rows and columns. Adding further, smaller squares result in adding errors and hence it is preferred to apply a Latin Square of more than 5*5.

3.5.6 Factorial Design

This type of design is used and preferred when multiple variables analysis is required to be done. In such cases, there exist different levels of the variables as well!!

Here the researcher is keen to understand the communications, interactions and impacts with the different variables.

For example: A researcher is interested in understanding the response/behaviour of people on particular days of week in the morning when they have breakfast and juice. A structure of the same is shown in Figure 3.10a. Two variables, breakfast and juice with different levels, are shown in Figure 3.10b.

	M	T	W	Th	F
Student 1	P	Q	R	S	T
Student 2	Q	R	S	T	P
Student 3	R	S	T	P	Q
Student 4	S	T	P	Q	R
Student 5	T	P	Q	R	S

FIGURE 3.9
A Design Arrangement for the Latin Square Example

	No juice	With juice
No breakfast	Response of participants without breakfast and juice	Response of participants without breakfast but with juice
With breakfast	Response of participants with breakfast but without juice	Response of participants with breakfast and with juice

FIGURE 3.10
(a) 2*2 Factorial Design

	No juice	With half glass of juice	With full glass of juice
No breakfast	Response of participants without breakfast and juice	Response of participants without breakfast but with half glass juice	Response of participants without breakfast but with one glass of juice
With breakfast	Response of participants with breakfast but without juice	Response of participants with breakfast and with half glass of juice	Response of participants with breakfast and with one glass of juice

FIGURE 3.10
(b) 2*3 Factorial Design with Levels

In factorial design, the number of levels in each factor decide the symmetry of the design. If for two factor the number of levels remain same, it is symmetric else asymmetric.

Figure 3.10 (a) Is a Symmetric Factorial Design Whereas. (b) Is Asymmetric.

Further a complex factorial design mentions the use when there are more than two variables involved. Generally, a researcher starts with a simple design and then proceeds in coming up with complex one. A complex design of 2*2*2 is shown in Figure 3.11.

Response from the different variables at different levels is captured with the complex factorial design. Breakfast here can be treated as the control variable, whereas the intake of juice made available are the different treatments been given to the participants.

Factorial designs show accuracy results at par with other designs and are an easy and economical way of obtaining the results. Further, the interactions and interest points can be captured with the graphs based on data values in the cells.

3.5.7 Quazi-Experimental Design

This type of design is used when the cost of experimentation is high. The design is used when the confounding relationships come in picture and it is not possible to control its impact on the experimentation. Also, there is no random assignment involved in this type of design.

The pre-test and post-test design we studied; fall under this category of experimental design. Time series, regression discontinuity are the other types of designs that are under this design type.

	With half glass of orange juice	With glass of orange juice	With half glass of pineapple juice	With glass of pineapple juice
No breakfast	Response 1	Response 2	Response 3	Response 4
With breakfast	Response 5	Response 6	Response 7	Response 8

FIGURE 3.11
Complex Factorial Design

Times series design – interrupted time series design is a variant of the pre-test and post-test design. In this type, the impacts are analysed at intervals with interruption. For example- assume that in a class, students' performance is to be measured for every weekly test that is conducted. To have interrupted time series analysis, an interrupt of some treatment takes place. This could be say scheduling group activities or some challenging session in between the weeks. Post this the performance is then measured. Unlike simple pre-test and post-test design where the impact analysis takes just once in between, this design is a series of pre-test post-test designs.

In case of Regression-Discontinuity (RD) design, it is a pre-test post-test design but a difference in the assignment. In the sense, what sets RD design different from others is the way the samples/participants are exposed and assigned to different treatments. A "Cut-off" score is calculated. This is on the basis of pre-experimentation, prior to the conduction of the design test. For example, to take decision regarding selection of a player in a team. In such cases a prior-performance cut-off can be employed and then the tests can be experimented. The issues here too are the assignment of pre-cut-offs and hence the dependency lies on the assignment.

3.5.8 Cross over Design

It is a design structure where the samples are supposed to undergo different treatments at different time intervals. This type of design belongs to clinical research or even can be applied in agricultural domain. Unlike parallel design where the sample is subjected to single treatment throughout, the crossover experiments with different treatment.

From medical perspective for example if a person (experimental sample) is undergoing clinical treatment; is treated with P treatment initially and then there is a washout/ wait over time period after which treatment Q is also introduced.

In such cases, the impact analysis is carried out at after every treatment. But there are difficulties while following this design as the impact of treatments could be negative.

Figure 3.12 shows the Cross over design pattern.

3.6 Sampling Concepts

Throughout the research design concepts in Section A, we mentioned about population, sample set, data sets and so on. How much important this is? Every experimentation needs large amount of data and to work with entire data available is difficult. Practically speaking, not necessarily the entire data is of importance and could contain some

Research Design

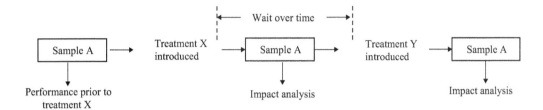

FIGURE 3.12
Cross over Design

redundant information. This could lead to biased opinions in the outcome for the impact analysis. A simple sample can be, for example, a group of 60 students from a population of 900. But are these 60 sufficient? Why only 60? On what basis? So, to what extent the samples taken influence the entire experimentation?

We will address these questions in this section.

Figure 3.13 depicts the sample and population relationship.

3.6.1 Design of Sample

We have studied research design where we discussed about the samples selected and assigned. Here we will discuss about the actual design of sample – the way in which the samples are obtained from the population. So, a method to capture, retrieve, select the samples from the entire population set that is available is referred to sampling.

3.6.2 Principles of Sampling

So, it is clear that we need to select particular and effective samples in sampling. The theory of sampling is based on two basic principles as follows:

(i) Principle of statistical regularity:

This principle is based on the probability theory. It states that a sufficient number of samples selected on random basis from the target population of study possess the required features of the population.

(ii) Principle of inertia of large numbers:
This principle states that larger the number of samples, more accurate the results would be.

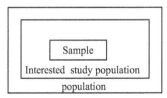

FIGURE 3.13
Sample-Population Relationship

3.6.3 Preparing a Sample Design

Prior to the selection and building of sample design, it is necessary to understand few aspects which are vital for preparation of the sample design.

(i) Population: The population is the entire group of participants a researcher would be looking at prior to carrying out the experimentation. This is essentially a large set. A population is a subset of the entire universe. So, to consider this, the researcher needs to understand the finite and infinite things of the universe. An example of population can be students in a school. This can be called as the target population, from which the interested study population is looked at. The target population needs careful selection. Many times, the researcher could take help of experts for this. Further depending on the population size, the interested study population and the sample is selected. If the population itself is very small in number, one might have to take it up entirely.

(ii) Sample frame: A sample frame contains list of data items with names from which the samples are to be taken. This is a very precise, comprehensive representation of the population and very much essential to prepare it.

(iii) Unit of sampling: What is it the researcher is interested in? A college, a factory, a housing society? This forms the unit. It could be a geographical area, or some locality.

(iv) Size of the sample: A most important and key factor that rules the experimentation impact. Often a point of concern to the researcher is "How many samples should I take?" A small sample size may not fulfil the requirements of the study resulting into lower accuracy results or wrong findings. But at the same time, taking large sample size result in generalisation and lack the actual precise findings. From this it is clear that samples that are suitable, adhering to required parameters of interest should be drawn. The experimentation outcome from these samples should be within the acceptable limits and the confidence intervals laid down.

(v) Parameters: In continuation to the sample size, the selection is governed by the parameters that the researcher knows are influential in the study. Consider a dataset of student. From this population, parameters like the address would be less important if the study is concern with the age groups of the students. Similarly, their mobile number, local address is of no use in the study if it is about their marks. But there could be some factors which influence. So, a meticulous selection of parameters is required.

(vi) Monetary requirements: A very common problem a researcher faces is the budget planning to get the sample. Collection of data say from medical field could be costly. This can impact the size of sample as well. A well-planned budget is necessary.

(vii) Method of sampling: Selection of techniques and methods to be applied for the sample selection are important. This represents the sample design. Considering the earlier-mentioned factors, the design selection must be able of accommodate and satisfy the requirements the researcher is interested in.

3.6.4 Sampling and Nonsampling Errors

In sampling theory, one of the factors that is important is the error. Sampling errors are the errors introduced owing to the inappropriate selection of the samples. That means the sample selected does not hold the potential to suffice the outcomes of the experimentation thus named as sampling error. So, the main reason for these errors is the deviation between the population and the sample means. (A sample mean is an estimate of the population mean.)

In the case of nonsampling errors, these are the ones that are formed due to human error. These errors can occur at every stage during the planning and the execution. Often happen when surveys are conducted. These errors can occur owing to absence of knowledge of the population, the problem to be solved, faulty approaches to the data collection and the data tabulation.

One thing is to be understood here that the data collection using surveys introduces both the types of errors. Also, one important point is that when the sample size is small, the sampling errors would be high and nonsampling errors would be less and vice versa.

A sampling design that would always have lesser sampling error should be the choice of the researcher.

3.6.5 Selection of Sampling Procedure

Though a researcher is keen on selection of a design that would minimize the sampling errors, there is a notion of "systemic bias" that occurs due to the sampling procedure that is selected. There is necessity to keep in control this bias. This comprises of sampling bias as well as nonsampling bias.

Few of the aspects for which the control needs to be established in order to reduce the impact of bias are discussed here.

(i) The sampling frame – selection of the frame for sampling affects the bias.
(ii) Equipment – the devices deployed to carry out the measurements also have a significant impact on the bias. Here it could be due to the human intervention who is responsible for surveys that is looked upon as device or it could be the actual physical equipment.
(iii) Non-participants – many times, people are reluctant to participate and give opinions in surveys or other data collections methods like interviews. In such cases, the inability to sample the participants results in bias.
(iv) Observable environment – data collection and sampling in an observed environment too adds up to bias. This is because the participants are aware that they are been observed and hence the efficiency will differ in comparison to an unobserved environment.
(v) Bias because of the publishing data – this kind of bias is the result of incorrect facts and figures been published. This is because where and who is going to publish the facts and figures is important. For an NGO, figures of poor, illiterate people would be large but for government organisations, they would portray it to be minimal.

So, to summarize, the sampling errors and the above-mentioned parameters need to be minimized and looked upon while selection of sampling procedure.

The steps in sampling process are:

(i) Identifying and defining the population that is to be experimented.
(ii) Defining the sampling frame.
(iii) Mentioning the sampling method, thus finalizing it.
(iv) Specifying the sample size.
(v) Employing the sampling plan as laid down along with selection of sampling units.
(vi) Actual sampling and data collection.
(vii) Reviewing the entire process.

3.7 Methods of Sampling

This section finally deals with the sampling designs – methods of sampling. Different designs are in place and as discussed, the overall research design and the outcomes are highly influential with the selection of the sampling design. Let us explore them.

3.7.1 Probability Sampling

In case of probability sampling, the samples selection is a fair decision. Each sample gets a fair chance of getting selected in the sample set. The probability of a sample getting selected can be determined. This type of design is also called as Equal Probability of Selection design. There are different variants of this sampling, let us study them.

(i) Simple random sampling

This type of simple random sampling technique is used when the population is small. Moreover, the population needs to homogeneous and easily available. In such circumstances, the sampling method gives every possible combination of the samples to get selected in the set with equal probability. There is a notion of sampling with replacement and without replacement.

Here for the random sampling, the selection of samples is without replacement. This means that once the sample is selected in the sample set, it will not be included again in the sample. So, just once. (But many times, formally, researchers would also be interested in sampling with replacement where the sample may occur more than once in the sample set.)

An example of this sampling can be random selection of students from a list of students studying in a college. A random number generator technique can be used to have the selection.

This sampling technique adheres to the principle of statistical regularity and hence many times is a preferred technique.

Research Design

(ii) Systematic sampling

In systemic sampling, it considers that the target population has some ordering and then goes into selection of every ith element at intervals.

Initially it begins with random number and then every ith element is selected from the population. For example, if i = 10; then initially from 1 to10 samples, the first sample would be taken random and then further every 10th sample will be included in the list.

Unlike the random sampling, here the spread and selection of sample is large. It can also be used in case of large sample size. One major drawback while using this type of sampling is that the samples selected at every ith interval could be biased or may be with some deformity. The technique is preferred when the population is easily available. But the results are not found to be reliable with this method. Figure 3.14 depicts the working of systemic sampling. Selection of ith sample, where i =10th sample.

(iii) Stratified sampling

In stratified sampling, the sample set constitutes of distinct categories of samples. The sampling technique is used when the population is heterogeneous. Because the population is nonhomogeneous, the sampling technique needs to ensure that the sample set comprises of every representative from the nonhomogeneous group or the distinct categories. Hence the population is split into different groups called as "strata." Each "strata" is more homogenous and bears common characteristics. The data items from these "strata" are sampled to form the sample set.

Figure 3.15 explains the stratified random sampling

There are a few things that need to be addressed while performing stratified sampling. The first one is deciding the formulation of strata. Is there any thumb rule? No. But the strata needs to be formed based on common features. This requires the researcher to know the data set and split it into the strata based on these features. Determining the variances among the population could also prove beneficial in the strata formulation.

The second thing that needs to be dealt is once the strata is formed, how to select the samples. Here simple random sampling can be applied. If required, system sampling too can be used.

The next important aspect is what should be the count of the samples selected? Once the population is split into different strata, the sample selection – the count from each strata should be in proportional to the count of the strata formed and the population.

To understand better, let us take an example. Suppose the population has 6,000 data items. Let this be called a P. From this P, we need to have sample set with count as n = 20. Assume that three strata are formed – S_1 = 3,000, S_2 = 1,800 and S_3= 1,200. Now from

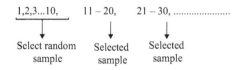

FIGURE 3.14
Systemic Sampling

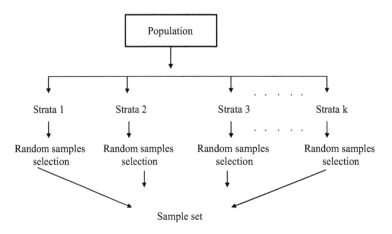

FIGURE 3.15
Stratified Random Sampling

these three strata, samples are to be selected in appropriate proportion to form the sample set of 20. So, the number of samples to be taken from each strata are given in the following equation:

$$SS_i = n^*(S_i/P) \qquad (3.1)$$

where SS_i represents the number of samples from "i"th strata selected.

n is the count of sample set, S_i is the number of samples existing in the "i"th strata and P is the population.

Putting the values for our example, we get
SS_1= 20 * (3,000/6,000) = 10.
SS2= 20 * (1,800/6,000) = 6.
SS_3 = 20 * (1,200/6,000) = 4.
So, in all the samples selected from each strata are 10,6 and 4 forming n = 20.

Cases in which the strata comprise of variable samples, the variability factor should be considered and then we have the revised equation to get the number of samples drawn from each strata as here:

$$SS_i = \frac{n^*(S_i\sigma_i)}{S_1\sigma_1 + S_2\sigma_2 + \ldots S_k\sigma_k} \qquad (3.2)$$

Where deviation factor of each strata is added. σ_i represents the standard deviation of ith strata. In all, k strata are formed.

In addition to the variability, the cost factor too can be considered. This is the cost of sampling of the strata. Following equation considers the cost factor as well:

$$SS_i = \frac{n^*(S_i\sigma_i)/\sqrt{C_i}}{S_1\sigma_1/\sqrt{C_1} + S_2\sigma_2/\sqrt{C_2} + \ldots S_k\sigma_k/\sqrt{C_k}} \qquad (3.3)$$

Where C_1 to C_k is the cost for sampling in strata 1 to k.

Research Design

A stratification sampling done with different characteristics is referred to as cross stratification. For example, a survey to understand the reviews of people working in IT industry with launch of new online car booking app can be with respect to the specific company, specific section of work in that company and so on.

(iv) Cluster sampling

In case of cluster sampling, the population is grouped to form relatively smaller groups or clusters. This sampling is used if the population is large enough and the entire area of population is divided into nonoverlapping groups or clusters.

So, for example if total population is of 6,000, smaller clusters – 120 can be formed, each comprising of 50 data items. Further, from these clusters, samples are randomly selected. Although there is no mention in literature about the number of n – samples to be drawn from each of the cluster. The sampling techniques proves to be economical and the impact analysis computed is sensibly accurate owing to the formulation of compact and precise clusters. It also helps in reduction of the cost as it deals with groups. Figure 3.16 shows the cluster sampling.

(v) Multistage sampling

This sampling procedure is an extended form of the cluster sampling. It comprises of levels of sampling from the clusters. For example, if an experimentation is to be performed to understand the use of social media on teenagers in schools in a state, the initial clusters could be on the basis of the talukas. Further from each taluka representative towns could be selected and then from them the schools and so on. This is multistage sampling.

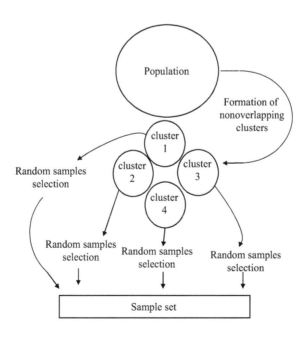

FIGURE 3.16
Cluster Sampling

If at every level, the selection of the samples is done at random, it is multistage random sampling. This type of sampling is used when the experimentation is involved with large geographical area. Owing to the sequential clustering, many units of samples can be accommodated in this type of sampling and also managing the clusters and gaining significant insights from clusters is also simple. There two reasons explain the use of multi-stage sampling. Figure 3.17 depicts the multistage random sampling.

3.7.2 Nonprobability Sampling

This sampling technique is not based on the probability of the sample selection. In fact, unlike probability-based techniques of sampling, where every sample had fair chance of getting selected in the sample set, the nonprobability sampling technique is the one where the researcher will select the samples. So, it is up to the researcher which samples to include. This sampling method is also known as deliberate or judgment sampling. The samples here loose the fair chance of making their way into the final sample set.

For example, if an analysis is to be carried out for faculty feedback in any institute regarding lecture conduction; the researcher would deliberately select the students who are most regular ones in attending. But many times this could lead to bias opinions as well. Let us explore different ways of nonprobability sampling.

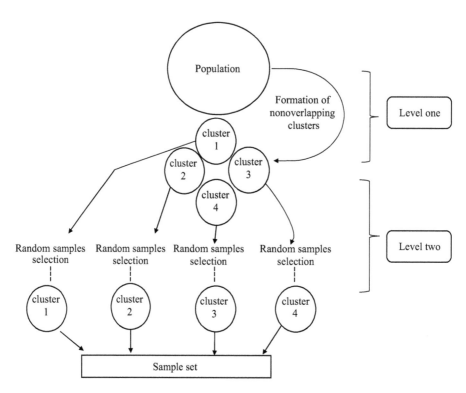

FIGURE 3.17
Multistage Random Sampling – Two Levels

(i) **Quota Sampling**

Somewhat related to the stratified sampling, this approach of quota sampling works on selection of representatives – quota but without probability. The population is split into subgroups and then based on judgments, the samples could be selected. For example, in an interview process, the HR personnel is responsible for selecting specific candidates from 150 candidates among which 80 are men and 70 are women (strata formed). The person has specific requirements of selection that 10% should be women from the total number of 45 vacancies to be recruited. The person may be biased with the selection for quota, where known candidates are likely to make way for the sample set. This is also again a type of judgmental sampling. So, everything is at the judgment and opinion of the interviewer.

(ii) **Convenience Sampling**

As the name suggests, this type of sampling relates in getting samples that the researcher finds convenient. This is nonprobability-based sampling and, at a given point in time, the researcher draws and selects samples readily available. For example, assume a survey needs to be conducted about hoteling on weekends in malls. Assume the researcher visits the mall and at that point of time only four to five participants are available. Inclusion of these samples will be biased and limited. This sampling lacks generalization and is preferred for pilot testing.

(iii) **Snowball Sampling**

This type of sampling technique is used when the researcher faces the problem of data collection and locating the samples. If the researcher is working on population involving tribal people, homeless people or any of such undocumented participants, there lies a great difficulty in reaching out and getting the data. In such cases, snowball sampling is used.

The researcher would generally locate some participant who is a part of the study and then rely on the information and sources from him to further locate participants. But such a sampling technique needs to include almost all the participants and often finds very few representatives of the samples as expected by the researcher.

3.7.3 Sampling in machine learning approaches

Till now we discussed about the different ways to capture the data samples and their applicability. While those are the most common methods, when it comes to machine learning approaches – percentile split, cross validation are the common approaches used to validate the results. Let us look at them.

Percentile split: A most common method that splits the available samples include labelled data (where the researcher is aware of the output class) into train data and test data. The train data is used to by the machine learning approach – a supervised one, to build a model. This model is then further tested with the test data. The percentage of samples to be used are generally 70% for train data and 30% for test data. However, in the case of noisy data, this could vary.

Cross validation: Often when we are dealing to build a robust classifier (a supervised machine learning approach), cross validation is used. This ensures good generalisation of the classifier and at the same time avoids overtraining. There are two variants of cross validation – 1. Hold-out and 2. K-fold

In case of hold-out technique for sampling the data, the entire labelled data available is split into three parts – (i) train, (ii) validate, and (iii) test. The classifier model is trained with the training subset samples and validated with the validated samples. The validated samples are used periodically to test the performance of the classifier. If the performance levels are acceptable, then test data is used to further validate the accuracy of the model generated.

K-fold cross validation is used to raise the performance higher than that of hold-out. The entire sample set – labelled one is divided into k parts, each bearing same size. One part forms the test data, remaining parts form the training data. A model is built based on the training data. and tested on the test data. The same process is repeated k times; each time with next part as test sample and all remaining as train samples. In each iteration, the performance of the model is noted and summarised after all the iterations are over. Often, the number – k is selected to be 10.

3.8 Summary

This chapter has focused on the necessity of appropriate and concise research plan to carry out the research without hindrances. The design explains the use and the ways in which the experimentation can be conducted and analysis of the impact can be carried out. The presence of extraneous variables and the problems to deal with it, are discussed in the chapter. Further, the sample selection and the techniques with which the researcher would work are extremely important to have good unbiased predictions of the experimentation.

Further Reading

1. C.R. Kothari, "Research Methodology – Methods and Techniques", New Age International Publishers.
2. Santosh Gupta, "Research Methodology and Statistical Techniques", Deep and Deep Publications.
3. Ranjit Kumar, "Research Methodology", Sage Publications.
4. Catherine Dawson, "Introduction to Research Methods", Viva books Pvt. Ltd.
5. https://onlinecourses.science.psu.edu/stat509/node/123
6. http://home.iitk.ac.in/~shalab/anova/chapter4-anova-experimental-design-analysis.pdf
7. https://wagner.nyu.edu/files/doctoral/Campbell_and_Stanley_Chapter_5.pdf
8. https://www.mff.cuni.cz/veda/konference/wds/proc/pdf10/WDS10_105_i1_Reitermanova.pdf

4

Basic Instrumentation

Pradeep B. Mane and Shobha S. Nikam

CONTENTS

- 4.1 Introduction 101
- 4.2 Characteristics of Instruments 101
  - 4.2.1 Static Characteristics 101
  - 4.2.2 Dynamic Characteristics....................................104
    - 4.2.2.1 Case Study-I: Sphygmomanometer 105
- 4.3 Instrumentation Schemes 106
  - 4.3.1 Block Diagram of Instrumentation Scheme 106
  - 4.3.2 Simple Instrumentation Scheme: Oxygen Cylinder 107
  - 4.3.3 Complex Instrumentation Scheme: Heat Exchanger..............108
- 4.4 Experimental Setup 110
  - 4.4.1 Laboratory Based Experimental Setup (Indoor Setup) 111
    - 4.4.1.1 Advantages 111
    - 4.4.1.2 Limitations..112
    - 4.4.1.3 Case Study for Laboratory Based Experimental Setup 112
  - 4.4.2 Field Based Experimental Setup (Outdoor Setup) 113
    - 4.4.2.1 Advantages 113
    - 4.4.2.2 Limitations..113
    - 4.4.2.3 Case Study for Field Based Experimental Setup...........113
  - 4.4.3 Natural/Quasi Experiments...................................114
    - 4.4.3.1 Advantages 114
    - 4.4.3.2 Limitations..114
    - 4.4.3.3 Case Study for Natural Experimental Setup 115
  - 4.4.4 Steps of Experimental Setup: Laboratory Experiments.........115
  - 4.4.5 Calibration of Instruments..................................115
    - 4.4.5.1 Need of Calibration 116
    - 4.4.5.2 Steps for Calibration...................................116
    - 4.4.5.3 Case Study: Calibration of Blood Pressure Measurement
             Equipment 116
- 4.5 Reliability of an Instrument 117
  - 4.5.1 Reliability Metrics 118
  - 4.5.2 Reliability and Availability 119
  - 4.5.3 Reliability Management.....................................119
  - 4.5.4 Reliability Characteristics.................................119
  - 4.5.5 Designing for Higher Reliability 125
  - 4.5.6 Reliability and Cost.......................................126
    - 4.5.6.1 Product Law of Reliability..............................127

4.5.7 Environmental and Other Physical Effects on Reliability 127
    4.5.7.1 Case Study 1: Medical Product Reliability Testing. 128
    4.5.7.2 Case Study 2: Temperature and Environmental Product Test 128
    4.5.7.3 Case Study 3: Reliability and Failure of Electrocardiogram (ECG) Machine 128
    4.5.7.4 Error Due to Lead Reversals 129
4.6 Data Collection 131
    4.6.1 Types of Data 131
        4.6.1.1 Advantages of Primary Data. 132
        4.6.1.2 Disadvantages of Primary Data 132
        4.6.1.3 Advantages of Secondary Data. 133
        4.6.1.4 Disadvantages of Secondary Data 133
    4.6.2 Sources of Data Collection 133
        4.6.2.1 Advantages 134
        4.6.2.2 Disadvantages. 134
        4.6.2.3 Advantages 135
        4.6.2.4 Disadvantages. 135
        4.6.2.5 Advantages 136
        4.6.2.5 Disadvantages. 136
4.7 Scaling 137
    4.7.1 Comparative Scaling 137
    4.7.2 Noncomparative Scaling 137
    4.7.3 Linear Scaling 139
        4.7.3.1 Limitations. 139
4.8 Role of Digital Signal Processing in Data Collection and Noise Filtering. 139
    4.8.1 Sources of Noise 140
    4.8.2 DSP in Noise Filtering 141
4.9 Measuring Instruments and Tools in Engineering Discipline 145
4.10 Summary 145
Further Reading. 159

> Always use branded items in life. For Lips – Truth, For Voice – Prayer, For Eyes – Sympathy, For Hands – Charity, For Heart – Love, For Face – Smile.
>
> —Abdul Kalam

Learning objectives of this chapter are to:

- Describe instrumentation schemes and identify process variables
- Distinguish static and dynamic characteristics and predict the reliability of an instrument
- Select experimental setup, test and verify hypothesis
- Choose appropriate data collection method
- Construct linear scale for data collection
- Apply Digital Signal Processing (DSP) for data filteration

This chapter will enable the researcher to:

- Understand scope of instrumentation schemes
- Recognize static and dynamic characteristics and predict the reliability of an instrument
- Categorize various factors improving reliability by reducing failures and errors
- Model experimental setups to prove hypothesis
- Describe major approaches in data collection
- Interpret the concept of linear scaling
- Employ data filtering on collected samples

4.1 Introduction

This chapter introduces the basics of instrumentation in general and discusses the characteristics and parameters of instruments that can be used by a researcher to perform experimental work to validate hypothesis formulated by him. Instrumentation facilitates interdisciplinary research. Researchers usually use instrument as a generic term for any measuring device (survey, test, questionnaire, and so on). To help distinguish between instrument and instrumentation, an instrument is a device and instrumentation is the course of action (the process of experimenting, testing, and using the device)[1]. The researcher should know the basic characteristics of any system/ instrument before choosing it for the experimentation. The following section discusses the static and dynamic characteristics of instrument. Researcher must select appropriate instrument to avoid inaccurate readings. For example, a typical lab 3½ Digital Multi Meter (DMM) having the limited frequency response (much below 10 KHz) cannot be used to measure the voltage at GSM frequency. Hence, researcher must know characteristics along with limitations of selected instrument for his/her research.

4.2 Characteristics of Instruments

It is of immense importance that, A researcher should understand basic system specifications and the generic characteristics of an instrument that appear in almost all specifications. There are two categories of performance characteristics of an instrument. They are mainly divided into[2]:

- Static characteristics
- Dynamic characteristics

Classification of performance characteristics of instruments (Figure 4.1):

4.2.1 Static Characteristics

When quantities being measured by instruments are either slowly varying with time or are constant, they are called as "static characteristics".

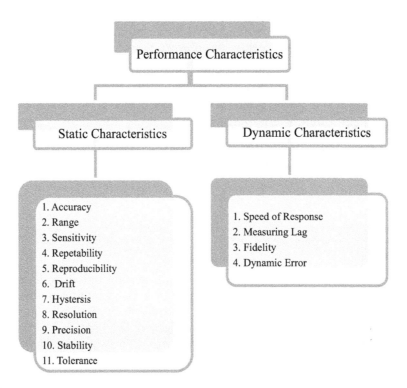

FIGURE 4.1
Classification of Performance Characteristics of Instruments[3]

(i) **Accuracy**: It is defined as the degree of closeness with which the measured value approaches the true value of the quantity being measured.

$$Accuracy = True\ Value - Measured\ Value \quad (4.1)$$

Accuracy is generally defined in terms of ± percentage. Usually the true value is unknown over all the operating conditions; so it is approximated with some standard. The accuracy is classified in the following ways:

- Point accuracy: It is specified at only one particular point of scale. Point accuracy does not give any information about the accuracy at any other point on the scale
- Accuracy as percentage of scale span: The accuracy of an instrument can be expressed in terms percentage of scale span when an instrument has uniform scale
- Accuracy as percentage of true value: True value is defined as an average of all measured values. Accuracy can be measured as percentage of true value over all measured values
- Precision: Precision is a measure of the degree of agreement within a group of measurements. It is a measure of the degree to which successive

measurements differ from one another. It is the measure of repeatability/reproducibility

(ii) **Range (or Span)**: The range/ span of an instrument is defined as the maximum and minimum values of the inputs or the outputs for which the instrument is recommended to use.

(iii) **Sensitivity**: It is defined as the ratio of the incremental output to the incremental input. While defining the sensitivity, it is assumed that the input-output characteristics of the instrument are approximately linear in that range.

(iv) **Repeatability**: It is defined as the variation in scale reading of an instrument for the same quantity to be measured.

(v) **Reproducibility**: It is defined as the degree of closeness with which a given value may be repeatedly measured by an instrument. It is specified in terms of scale readings over a given period of time.

(vi) **Drift**: Drift is grouped into following three categories:
- Zero drift: Zero drift occurs in an instrument when whole calibration gradually shifts due to undue warming up of an electronic component, slippage and permanent set
- Span drift or Sensitivity drift: If there is proportional change in the indication all along the upward scale, the drift is called span drift or sensitivity drift
- Zonal drift: It is defined as drift for a portion of span of an instrument

The details of drift (span and zero) are depicted in Figure 4.2.

(vii) **Hysteresis**: Hysteresis exists not only in magnetic circuits, but in instruments also. For example, the deflection of a diaphragm type pressure gauge may be different for the same pressure, but one for increasing and other for decreasing. The hysteresis is expressed as the maximum hysteresis as a full-scale reading[4].

(viii) **Resolution**: Resolution is defined as the smallest change in input which can be measured by the instrument. If the input is slowly increased from some arbitrary input value instrument does not respond until a certain increment is exceeded. This increment is called as resolution as depicted in Figure 4.3.

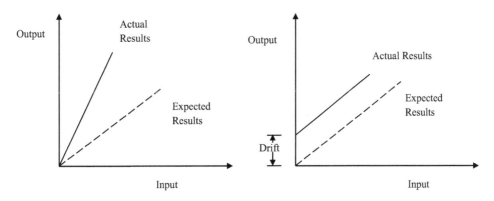

FIGURE 4.2
Span Drift and Zero Drift

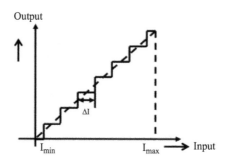

FIGURE 4.3
Resolution

(ix) **Precision**: A researcher should understand difference between precision and accuracy. Precision indicates the repeatability or reproducibility of an instrument, but does not indicate accuracy. If an instrument is used to measure the same input, but at different instants of time, the successive measurements may vary randomly. The random fluctuations of readings are commonly represented with a Gaussian distribution. A precision of an instrument indicates that the successive reading would be very close, or in other words, the standard deviation σ_e of the set of measurements would be very small.

(x) **Stability**: It is defined as the ability of an instrument to maintain its performance throughout its specified operating life.

(xi) **Tolerance**: Tolerance is defined as the maximum allowable error in the measurement of an instrument.

4.2.2 Dynamic Characteristics

When quantities being measured by instruments changes rapidly with time, they are called as "dynamic characteristics."

The dynamic characteristics are categorized as follows[5]:

(i) **Speed of response**: It indicates how fast the measuring system responds to changes in the measured quantity.

(ii) **Measuring lag**: It is the retardation or delay in the response of a measurement system to changes in the measured quantity. The measuring lags can be categorized as follows

- Retardation type: In this type of measuring lag an instrument responds immediately after the change in quantity being measured occurs
- Time delay lag: If the response of an instrument begins after a dead time on the application of the input this type of measuring lag is called as time delay lag

(iii) **Fidelity**: It represents faithful reproduction. It is the degree to which an instrument indicates the changes in the measured variable without dynamic error.

Basic Instrumentation

(iv) **Dynamic error**: It is defined as the difference between true value of the quantity being measured changing with time and the value indicated by the measurement system assuming zero static error. It is also called as measurement error.

Performance of the instrument depends on its static and dynamic characteristics. It allows user to select the most suitable instrument for a specific measuring job. Dynamic behavior and steady state behavior of the system is closely related to its performance characteristics. In the following subsequent topic, the case study of sphygmomanometer is studied to understand the static and dynamic characteristics of the instruments/systems.

4.2.2.1 Case Study-I: Sphygmomanometer

Medical equipment is designed to assist in the diagnosis, monitoring or treatment of medical conditions. To understand these explained characteristics, let us consider the Noninvasive (or Indirect type) Sphygmomanometer Blood Pressure Measuring equipment (depicted in Figure 4.4).

(i) **Accuracy**:

In the fields of science and engineering, the accuracy of a measurement system is the degree of closeness of measurements of a quantity to that quantity's true value. Standard range of Blood Pressure is 80 mmHg-120 mmHg if measured range is 79 mm Hg-119 mm Hg then it can be concluded that accuracy of an instrument is ± percentage of standard reading. Accuracy = ± (0.1% of Range ± 1 mmHg).

(ii) **Span**:

As illustrated in Figure 4.4 for a Noninvasive type (Indirect type) Sphygmomanometer Blood Pressure Measurement instrument the minimum or start reading

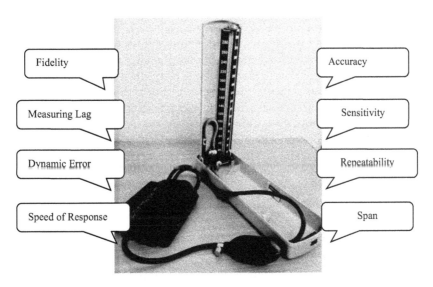

FIGURE 4.4
Sphygmomanometer

range is 0 mm Hg and maximum or end reading is 240 mmHg. It can be said that span of instrument is 0 mmHg – 240 mmHg.

Dynamic Characteristics:

(i) **Fidelity**: The system is said to have fidelity if it reproduces output same as the input. E.g. Blood pressure will be 80 mm Hg-120 mm Hg and tested equipment reading is also 80 mm Hg-120 mm Hg then it can be said that test equipment is having 100% Fidelity.

(ii) **Speed of Response**: Speed of response is the time required for a system to react to some signal and to indicate the output. In digital type blood pressure measurement instrument, after providing the input, time taken to show the systolic and diastolic pressure reading output is called as speed of response.

(iii) **Measuring Lag**: Time taken by Sphygmomanometer to respond to the changes in input.

(iv) **Dynamic Error**: Let us consider the condition at the time of calibration, standard reading at normal stage is 80 mm Hg-120 mm Hg but the instrument indicates 70 mm Hg-110 mm Hg while taking the readings. Then it can be said that 10 mm Hg difference in Systolic and Diastolic reading is the dynamic error.

(v) **Reliability**: It gives validity and consistency of sphygmomanometer. Test results are determined through statistical methods after repeated trials. For example patient's first reading is 80 mm Hg-120 mm Hg and if the readings remain same for repeated trials over a span of time then the instrument is said to be reliable.

4.3 Instrumentation Schemes

An instrumentation scheme comprises two words, one is Instrumentation and other is scheme. The former is art and science of measurement and control of the process variables within a production or manufacturing area; whereas the later means a systematic plan or arrangement for attaining some particular objective or putting particular ideas into effect. So, combining these two, we can say that the Instrumentation scheme is a systematic arrangement of various instruments to attain desired objective. Generally, in any industry, the process variables used are; Level, Temperature, Pressure, Humidity, Flow, pH, Force, Speed, and so on. Instrumentation scheme identifies any of the earlier-mentioned process variables and converts it into some usable form, that is, electrical form. This electrical form of physical variable can be further used either to display for monitoring or control or in some interrelated processes to attain desired output. The general block diagram of instrumentation scheme is discussed in the following section.

4.3.1 Block Diagram of Instrumentation Scheme

The instrumentation scheme can be either simple or complex depending on the application and the specifications of the system. The instrumentation scheme for a process, a system, or an application includes following blocks as illustrated in Figure 4.5.

Basic Instrumentation

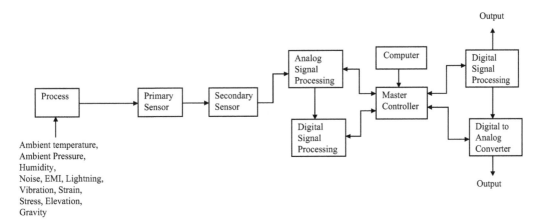

FIGURE 4.5
General Block Diagram of Instrumentation Scheme

The environment where process/experiment is performed can affect the process/experiment, measuring sensors and equipment thereby disturbing measurement results. Variables such as ambient temperature, pressure, noise, EMI, stress, strain, vibrations, gravity, and elevation can impact the measurement results. The primary sensor senses the appropriate physical variable. If necessary, secondary sensor comes in picture before converting physical parameter into electrical form. This electrical signal is then transferred to master controller through analog signal processing and analog to digital convertor units. Master controller then takes appropriate corrective or controlling actions and sends corrected digital signals to digital signal processing and digital to analog convertor unit. The corrected analog signal is then sent to final control element to get desired output from the process.

4.3.2 Simple Instrumentation Scheme: Oxygen Cylinder

Oxygen cylinder is an oxygen storage vessel, which is either held under pressures in gas cylinder, or as liquid oxygen in cryogenic storage tank. Nowadays, oxygen cylinder are used in variety of applications like in hospitals as a breathing gas, in rocket engines as liquid rocket propellants, during mountain climbing and scuba diving, and so on. As oxygen is very much prone to fire, its pressure is rarely held higher than 200 bar/3000 psi. Safety considerations limit the upper value while for normal operation the pressure needs to be above a certain value. Apart from restricting the pressure within these two limits, rarely does the need arise to measure or regulate it precisely. Oxygen cylinder is normally fitted with pressure gauges that indicate the contents of the gases in the cylinder. The range of the pressure gauge is from 0 to 200 kg/cm^2. Whenever the new oxygen cylinder is hooked up and taken in line, the pressure indicator should be above this mark to confirm the cylinder is filled. With the gradual usage of this cylinder the reading would drop gradually, when the indicators show that the pressure has fallen below the minimum level of acceptance, the cylinder should be refilled. On the other hand, if for any reason, the pressure gauge shows the reading above 200 kg/cm^2 during use, the cylinder should be disconnected immediately and replaced. The accuracy of pressure indicator should be ±1% of range span. Here the basic function of pressure gauge is to indicate oxygen cylinder pressure. The minimum requirement of this instrument (pressure gauge) is, it should be robust, accurate, and reliable to avoid accidents. Figure 4.6 shows the simple instrumentation scheme for oxygen cylinder.

4.3.3 Complex Instrumentation Scheme: Heat Exchanger

The complexity of the system increases with the increasing number of measuring and controlling parameters. Consider a heat exchanger process as depicted in Figure 4.7 with multiple measuring and controlling parameters. Heat exchanger is a means where heat can be exchanged between solid objects and liquid or between two or more liquids. To prevent direct contact or mixing of these liquids, it holds solid walls as a partition. The commonly used type of heat exchanger is a shell and tube type as it is suitable for wide range of pressure and temperature applications.

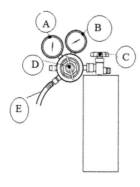

A: Bottle Pressure Gauge
B: Output Pressure Gauge
C: Oxygen Bottle Valve
D: Regulator Control Valve
E: Oxygen Hose

FIGURE 4.6
Oxygen Cylinder Instrumentation Scheme

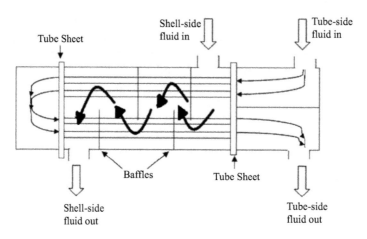

FIGURE 4.7
Shell and Tube Type Heat Exchanger

Basic Instrumentation

It consists of an outer shell with tubes inside it. The process fluid whose temperature is to be controlled flows through the shell and other fluid like steam flows inside the tubes to exchange the heat between the two fluids. The more is the steam flowing through tubes the more amount of heat can be exchanged and vice versa.

Figure 4.8 shows the control loop for heat exchanger

Here, the temperature of process fluid OUT shown as TTI1 OUT is depending upon various parameters like

- Temperature of process fluid IN (TTI1 IN)
- Flow rate of process fluid IN (FTI1)
- Flow rate of steam IN (FTI2)

So, to take process fluid OUT temperature at desired value (set point), it is necessary to keep all above parameters in the suitable range. To achieve this, the process fluid IN line, OUT line and steam line should have appropriate temperature and flow rate indicators and respective transmitters. Indicators display process parameter value locally while transmitters convert that process parameter value in suitable electrical form. This electrical form is then transferred to controller to take controlled actions. By implementing appropriate control algorithm, controller drives the process fluid IN temperature, fluid IN flow rate and steam IN flow rate in suitable range by sending controlled signal to respective final control element.

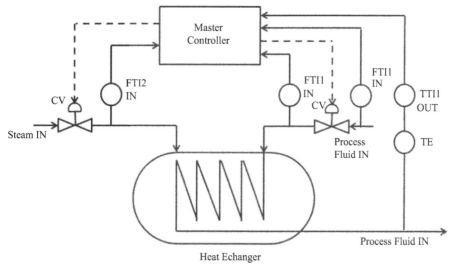

TE: Temperature element
TTI1 OUT: Process fluid OUT temperature Transmitter & Indicator
TTI1 IN: Process fluid IN temperature Transmitter & Indicator
FTI1 IN: Process fluid IN flow transmitter & Indicator
FTI2 IN: Stream IN flow transmitter & Indicator
CV: Control Valve

FIGURE 4.8
Heat Exchanger Control Loop

To make this heat exchanger more sophisticated we can add personal computer (PC) in the loop so that:

- The set point can be changed with requirements
- Commands to start and stop the process can also be given from keyboard

The overall instrumentation and control scheme for this heat exchanger is shown in Figure 4.9.

4.4 Experimental Setup

A Researcher formulates hypothesis to solve the problem or to bridge the gap in existing system using mathematical model or scientific methods. In order to test hypothesis, a researcher performs experiments. Experiment is divided into two groups, experimental groups and control groups. It provides a standard of comparison to examine if any change occurs as a result of independent variable. The experimental group gets the independent variable. The control group does not get the variable, it has dependent variables. It will provide platform for the data analysis and the researcher may accept or reject the hypothesis.

An experimental setup is systematic and scientific approach used to test the effect of variables on an outcome to support or validate a hypothesis. The variables that the researcher manipulates are called as the independent variables. The variables which are

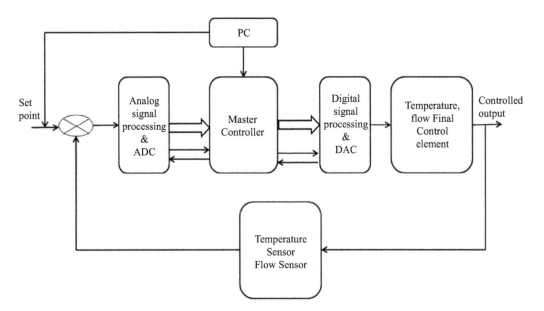

FIGURE 4.9
Instrumentation Scheme for Heat Exchanger

Basic Instrumentation

controlled and/or measured are dependent variables. It plays an important role in the development of a researcher, as it develops continuous learning attitude of a researcher. In science, the experimental setup is the part of research. Broadly experimental setup can be divided into three types:

- Laboratory based,
- Field Based,
- Natural Experiments

4.4.1 Laboratory Based Experimental Setup (Indoor Setup)

A laboratory provides controlled conditions in which experiments, measurements or research can be performed. Those experiments, measurements, or research work that have laboratory-based setup enable the experimenter to measure the effects of independent variables on dependent variables precisely, thus establishing cause and effect relationships. With this, they can predict the behavior of dependent variable in the future. Various fields of science and engineering have different requirements. As per those requirements laboratory setups take many forms. For example, Physics, Metallurgy, Chemist, Biologists, Computer Scientist, and Technical Institute Laboratories have various equipments, apparatus, and laboratory conditions.

In technical institutions, traditional real laboratories are used, but they have lot of limitations. The basic as well as advanced laboratory equipment required for research may be limited due to lack in availability of resources. The limited working hours of an institute may restrict the researchers from doing experiments according to their own schedules. With the advancement and modernization in technology for engineering education, it is possible to access the remote laboratory setup via web, such setups are called as Web-based experimental setup or virtual laboratory experimental setup. Use of virtual laboratories for performing experiments is one of the most efficient ways of utilizing available technical resources.

4.4.1.1 Advantages

- Control – laboratory experiments have a high degree of control over the environment and other extraneous variables, so experimenter can assess the effects precisely. It results in enhanced accuracy of results. Also, it has higher internal validity
- Isolation of variables – controlled condition of laboratory experiment eliminates the effect of extraneous variables. These variables are the undesirable variables like noise which are not of interest but might interfere with the results of the experiment. Also, a controlled condition of laboratory experiment can isolate independent variables easily than other experiment methods. So, it becomes possible for an experimenter to test results with one or more independent variables easily
- Reliability – Due to high levels of control, experiments can be performed repeatedly, so that the reliability, precision of results can be checked

4.4.1.2 Limitations

Lack of external validity: As lab experiments are conducted in controlled condition, that is, artificial environment, these conditions are far different from natural real-life conditions, the results tell very little about how dependent variables will act in real life conditions, that is, the results cannot be easily generalized. Due to this limitation, sociologists hardly ever use lab experiments.

4.4.1.3 Case Study for Laboratory Based Experimental Setup

Testing of the antenna parameters in anechoic chambers is discussed below as a case study of laboratory based experimental setup.

The laboratory based experimental setup to test antenna parameters includes anechoic chamber. An anechoic chamber is a huge Faraday enclosure, which eliminates reflection, external noise and interference caused by electromagnetic waves. All inner surfaces of the chamber are lined with RF/microwave absorbers, which minimize the amplitude and phase ripples in the test zone. Hence, an anechoic chamber provides accurate and precise measurement of antenna parameters such as antenna far-field patterns, gain, directivity, radiation efficiency, input impedance and polarization. It provides controlled electromagnetic environment. A picture of an anechoic chamber is shown in the Figure 4.10 along with experimental setup to measure antenna parameters.

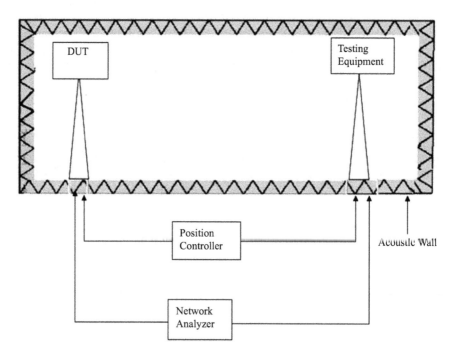

FIGURE 4.10
Anechoic Chamber to Test Antenna Parameters

4.4.2 Field Based Experimental Setup (Outdoor Setup)

Some experiments need to be performed in the real world, that is, at a field place rather than in the laboratory, are known as field-based experiments. Field experimental setups are completely different from laboratory experimental setup. Laboratory experiments enforce scientific control by testing a hypothesis in the artificial and highly controlled setting of a laboratory. These experiments take place in real-life settings such as industries, geographical locations, classroom, workplaces, and so on. The field experiments are sometimes seen as having higher external validity than laboratory experiments as they are performed in natural environment rather than in a laboratory environment. However, like natural experiments, field experiments suffer from the possibility of contamination: experimental conditions can be controlled with more precision and certainty in the laboratory.

4.4.2.1 Advantages

- External validity – As field experiments take place in normally occurring natural conditions, they have reasonable external validity than lab experiments
- Larger-scale involvement – As it is possible to do field experiments in large premises like in large institutions, in schools or workplaces, large number of people can interact in the process

4.4.2.2 Limitations

- Less control: In field experiments, experimenter has less control on variables than lab experiments which might result in more distortion of findings due to interference of extraneous variables and which lead to lower internal validity
- Experimental effect: Field experiments are generally performed with larger scale settings involving more people who may reduce the validity of results as respondents may act differently just because they know they are part of an experiment
- Practical problems: For lab experiments, access is not a major issue, but with field experiments, and workplaces, institutions might not be willing to allow experimenter to do so

4.4.2.3 Case Study for Field Based Experimental Setup

In some applications like an antenna mounted on an aircraft, mobile systems parameters cannot be tested in Laboratory (anechoic chambers) hence, field based experimental setups are required.

The antenna parameters that can be tested in outdoor ranges are far-field patterns, gain, directivity, radiation efficiency, input impedance and polarization. These parameters are affected by Electromagnetic environment and by other RF losses like fading, free space path loss, ground reflections, shadowing, and so on. The outdoor sites for antenna parameter measurement provides propagation of signal in the line of sight (LOS) as illustrated in Figure 4.11 which must be unobstructed and free from ground and other reflections from nearby objects like human beings, buildings, ground, and so on. The reflected signal in elevated range is the major cause for error in radiation pattern and gain measurements.

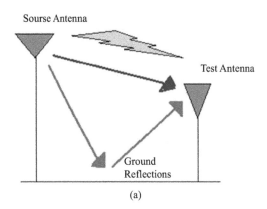 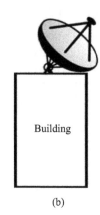

FIGURE 4.11
Illustration of Outdoor Range Setup

The researcher must take care while performing experiments on field (outdoor) as the environmental effects usually degrade the measured values.

4.4.3 Natural/Quasi Experiments

In natural experiments, experimenter does not manipulate variables deliberately as the variables change naturally. So, experimenter just measures the effect of something that is already happening. Natural experiments are used in observational studies. Natural experiments are employed as study designs when controlled experimentation is extremely difficult to implement or unethical example evaluating health impact due to particular disease. When controlled experimentation like Laboratory experiment is not possible to study the situation, in such cases, Natural experiments are often performed. Situations like policy changes, weather events and natural calamities create circumstances for natural experiments. Natural experiments are used most commonly in the fields of epidemiology, political science, psychology, and social science.

4.4.3.1 Advantages

High external validity – As experimenter does not manipulate the variables and variables change naturally, the findings/results can be easily generalized to real-life settings. So, these experiments have high external validity.

4.4.3.2 Limitations

- Lack of control – In these experiments, as variables change naturally and experimenter does not have control over them, extraneous variables and other environmental settings can affect findings. So, these experiments have low internal validity
- No repetition – Experiments cannot be repeated as the researcher has no control over variables. Reliability of results cannot be checked in these experiments

4.4.3.3 Case Study for Natural Experimental Setup

Effect of Smoking on Health is discussed below as a case study of natural experimental setup.

In Helena, Montana a natural experiment regarding smoking ban from all public places was performed over a period of six-month. A study was carried out on health issues. When the ban was in effect, 60% drop in heart attacks was reported by investigators for the study area. As such experiments cannot be performed in the Laboratory, they fall under the category of natural experiments [6].

4.4.4 Steps of Experimental Setup: Laboratory Experiments

Figure 4.12 indicates the steps of experimental setup as detailed below:

(i) **Identify the Problem Statement**: Initially the researcher has to define the Problem statement, which helps him/her to study in more narrow area appropriately. Defining proper problem statement leads in formulation of proper hypothesis.

(ii) **Formulate the Hypothesis**: Hypothesis is the objective of experimental setup. It gives direction to the researcher. In a hypothesis statement, the researcher makes a prediction about what he/she thinks will happen in the experiment. They predict about the results.

(iii) **Test the Hypothesis with experimentation**: To test a hypothesis, the researcher performs experimentation on input data. While conducting experiment, they have to identify and control the non-experimental factors like noise which can affect the result. Selection of instrument should be proper. Special focus should be given on accuracy, precision, range, response time of instrument. Instrument should be calibrated before use. A Researcher can perform an experiment with properly selected instrument. After experimentation they have to collect and compile the obtained results in a systematic way that can be usable.

(iv) **Analysis of Experimental results**: After getting output i.e. result they check the obtained result with the actual one. This is analysis of data. If the obtained result is not within acceptable range as per the hypothesis, the above steps are repeated. The above process is repeated until results are acceptable. In some cases, the researcher may have to make some modification in hypothesis too.

(v) **Conclusion**: Finally with the help of analyzed data researchers formulate the conclusion. In the conclusion researcher has to comment on observations, they examined through-out the experiment process. Repetition of experimentation is necessary because when same results are obtained, it gives confidence to results.

To test hypothesis researcher collects data and then performs experimentation on collected data using appropriate instrument. It is of prime importance that researcher should select instrument which will show true value at the output, and for that calibration of instrument with standard instrument is necessary.

4.4.5 Calibration of Instruments

Calibration is a method used to compare the measuring, and test instruments to a recognized reference standard of known certified accuracy and precision, noting the difference and adjusting the instrument, where possible, to agree with the standard.

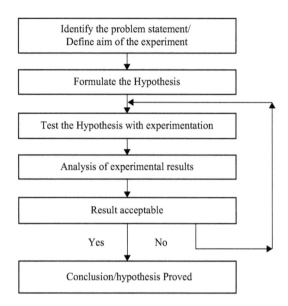

FIGURE 4.12
Flow Diagram Indicating the Steps of Experimental Setup

4.4.5.1 Need of Calibration

Instrument error can occur due to a variety of factors: drift, environment, electrical supply, addition of components to the output loop, process changes, and so on. Since calibration is performed by comparing or applying a known signal to the instrument under test, errors are detected by performing calibration.

4.4.5.2 Steps for Calibration

Step 1: Identify the Measuring Devices/Instruments
Step 2: Determine Certification, Calibration, and Accuracy Check Requirements
Step 3: Calibration process
Step 4: Corrective Action
Step 5: Verification
Step 6: Documentation and Record Keeping

4.4.5.3 Case Study: Calibration of Blood Pressure Measurement Equipment

The steps to calibrate self-measured blood pressure device at doctor's clinic to measure its accuracy against a mercury sphygmomanometer or primary device are discussed below.

1. Make the subject under test (Patient) sit down with his/her arm at heart level completely relaxed.
2. Before starting the measurement ask the patient to rest/relax for five minutes.

3. During the measurements do not allow the subject under test (Patient) to converse with anybody to prevent change in blood pressure.
4. With a gap of 30 seconds take a total of five sequential same-arm blood pressure readings.
5. Take the first two readings with the BP device under calibration.
6. Take the third reading, preferably with a mercury sphygmomanometer or primary device by the health care clinician.
7. Take again the fourth reading with the same BP device under calibration.
8. Take the fifth and final reading again with a mercury sphygmomanometer or primary device by the health care clinician.
9. Compare the difference between the readings from the two cuffs.
    a. BP readings generally reduce over the five measurements. The final systolic blood pressure reading may be as much as 10 mm Hg lower than the first.
    b. The comparison is acceptable if the difference is 5 mm Hg or less.
    c. If the difference is greater than 5 mm Hg but less than 10 mm Hg. Repeat the above calibration procedure.
    d. If the difference is greater than 10 mm Hg the device may not be accurate.
10. Repeat this procedure if the device is not calibrated as per the specifications.

4.5 Reliability of an Instrument

Any system/instrument cannot work satisfactorily over a long period of time unless the failure rate of the components is zero. In practice it is not possible to manufacture each component perfectly as some other factors will also force it towards failure. Component failure results in system/instrument failure. To avoid this, component is manufactured with greater reliability (generally R = above 0.9899 as a standard). In order to increase the reliability of an instrument/system the concept of redundancy can also be used which will help to avoid stoppage of work due to system failure.

A failure is partial or total depending upon the loss or change in the properties of an instrument in such a way that functioning is partially or completely stopped. Failure, in general can be grouped into different modes depending upon nature of failure which can be categorized as below:-

- Initial failure: Occurring during initiation phase
- Early failure: Occurring after a short time after installation
- Random failure: Occurring at an arbitrary instance
- Catastrophic failure: Occurring due to major failures and stopping the
- functioning for long time
- Wear out: Occurring after long usage of system/instrument
- Failure density: This is the ratio of number of failures during the given unit of time interval to the total number of items at the beginning of the test. The failure

density is usually carried out during the trail phase of any system/instrument to know the performance parameters
- Failure rate: It is defined as ratio of number of failures to number of component hours. Hence for electronic components Failure rate (FR) is generally stated as:

$$Failure\ rate(\lambda) = \frac{Number\ of\ failures}{Number\ of\ component\ hours} \qquad (4.2)$$

Failure rate is usually a calculation done when the system/instrument is used for a long duration of time to know its performance and state.

Further the key causes of failure and unreliability are:

(i) Poor design
(ii) Wrong manufacturing techniques
(iii) Lack of total knowledge and experience.
(iv) Complexity of system/instrument.
(v) Poor maintenance policies
(vi) Bad quality control
(vii) Storage or transport
(viii) Misuse and mishandling
(ix) Human errors (Human errors can be further classified)
- Lack of knowledge of equipment
- Lack of understanding process
- Carelessness
- Forgetfulness
- Poor judgment skills
- Physical inability

4.5.1 Reliability Metrics

Reliability is the ability of the instrument/system to maintain its quantity and quality under specified condition for a specified time. Reliability of a system/instrument is the probability that the system/instrument performs intended function adequately for given period of the time under the stated operating condition or environment. The reliability definition stresses four elements namely:

- Probability
- Intended function
- Time
- Operating Conditions

There is a thin line of difference between reliability and quality. It goes hand in hand. In conclusion; there cannot be reliability if the system /instruments are not of quality. The Comparison between Reliability and Quality is given in Table 4.1.

4.5.2 Reliability and Availability

Although the definition of availability appears to be similar to reliability, availability has different meaning. Reliability emphasis on failure free operation up to time t whereas availability means the status of equipment at time t. The availability function does convey about the number of failure that occurs during time t, but depends upon both the failure and repair rates.

4.5.3 Reliability Management

Reliability management is concerned with performance and conformance of the system/ instruments over the expected life. Reliability management includes planning, designing, verifying and tracking of the products throughout their life to achieve reliability. Reliability of a system is often specified by the failure rate λ. For most of the systems/ instruments, $\lambda(t)$ is a "bath-tub curve": To do the reliability management, reliability characteristics play a very important role.

4.5.4 Reliability Characteristics

The reliability characteristics are classified as Repairable Systems and Non-Repairable Systems.

(i) **Parameters playing vital role for Non-Repairable Systems**:
- Reliability=Availability
- Failure Rate: The number of failures per unit of gross operating period in terms of time, events, cycles
- MTTF (Mean Time to Failure): It is the average time to first failure occurrence

TABLE 4.1

Comparison between Reliability and Quality

Reliability	Quality
Reliability of an instrument/system is the probability that it performs its intended function adequately for a given period of time under stated operating condition or environment.	Quality of instrument/system is the degree of conformance to applicable specification and workmanship standards.
Reliability is primarily associated with designing of a system /instruments.	Quality is associated with manufacturing of a system /instruments
Reliability depends on time and environmental condition	Quality is not concern with time and environment condition
Reliability can be defined by numerical value.	Quality cannot be defined by numerical value
For reliability creating a redundancy in the system is mandatory for better reliability and avoidance to failure rate.	It is impossible to construct good quality system by having a redundancy.

- MTTF is a statistical value and is meant to be the mean over a long period of time and
- a large number of units
- Time to First Failure: The first failure occurrence during the operating period
- MRL (Mean Residual or remaining Life): It is the mean time for which the system/instruments works after which the failures has already occurred and before stoppage

(ii) **Parameters playing vital role for Repairable Systems:**
- Availability (Function of Reliability and Maintainability)
- Failure Rate and Repair: It is defined as ratio of number of failures to number
- of component hours and the repairs done after the occurrence of the same
- MTBF (Mean Time Between Failures): It is the average time between failure
- occurrences. The number of items and their operating time divided by the total
- number of failures for repairable items
- MRL(Mean Residual or remaining Life): It is the mean time for which the
- system/instruments works after which the failures has already occurred and the
- maintenance and repairs costing is taken into consideration

To calculate reliability for any system/instrument some important formulas are stated for ready reference:

- $$\text{MTBF}(\theta) = \text{Total time}/ \text{Total Number of failures} \qquad (4.3)$$

- Average Failure Rate

$$\lambda = 1/\theta \rightarrow \lambda\theta = 1 \qquad (4.4)$$

- Hence λ= Total Number of failures/ Total time
  Mean time to failure (MTTF)

$$= 1/\lambda \qquad (4.5)$$

- Maintainability can be measured with Mean Time to Repair (MTTR), MTTR is average repair time and is given by:

$$\text{MTTR} = \frac{\text{Total Maintenance Down Time}}{\text{Total Number Of Maintenance Actions}} \qquad (4.6)$$

- Failure Rate:

$$= \frac{f(t)}{R(t)} \qquad (4.7)$$

Example 4–1. 600 buses have accumulated for 54,000 hours, 10 failures are observed. What is the MTBF? What is the failure rate?

Solution: Consider car as repairable system,
MTBF(θ) = Total time/Total Number of failures
MTBF = 54,000/10 = 5400 hours.
Average Failure Rate $\lambda = 1/\theta$
Average Failure rate $\lambda = 10/54,000 = 0.000185$ per hour.

Example 4–2. Six water pumps were tested with failure hours of 54, 32, 52, 96 and 104. What is the MTTF and failure rate?

Solution: consider pumps as nonrepairable systems

$$\text{MTTR} = \frac{\text{Total Maintenance Down Time}}{\text{Total Number of Maintenance Actions}}$$

MTTR = (54+32+52+96+104)/6 = 56.3 hours
Failure rate $\lambda = 6/(54+32+52+96+104) = 0.01775$ per hour.
Note that MTTF is a reciprocal of failure rate, so MTTF = $1/\lambda$
Substituting the values of λ,
MTTF is 56.33 hours

Example 4–3. In an illustration here assume 10 components were tested and for Component1 to 5 failed after 58, 26,132,323,520 hours. Component 6 to 10 failed after 300,400,525,700,400 hours. The components failed are not repairable. Find the failure rate and mean time till failure. What is the mean time till failure assuming a constant failure rate?

Solution: Calculate failure rates from historical information.
No. of failures = 10.
Total operating time as per the data furnished = additions of all failure rates
58 + 26 + 132 + 323 + 520 + 300 + 400 + 525 + 700 + 400 = 3384 hours
Failure rate is calculated as
λ = No of failures/total operating time
$\lambda = 10/3384 = 0.00295$
Mean time till failure (MTTF) = $1/\lambda = 1/0.00295 = 338.40$ hours.

Example 4–4. For example, if you had 6 hydraulic pumps in standby mode and each ran for 1000 hours in standby and 4 failed during the standby time. Find the failure rate in standby mode.

Solution: Total standby hours = No. of pumps * standby time
= 6*1000 hours = 6000 hours
Failure rate in standby mode = No. of pumps Failed/Standby time
= 4/6000
= 0.00066 failures per hour

The other Important Analytical Functions in Reliability Engineering are:

- Failure Probability Density Function
- Failure Rate Function
- Reliability Function

- Conditional Reliability Function
- Mean Life Function.

Reliability Function and Failure Rate Calculations:

For a probability density function $f(x)$ for the time till failure, reliability functions can be defined as;

$$R(t) = P(T>t) = \int_t^\infty f(t)dt = 1 - \int_0^t f(t)dt = 1 - F(t) \qquad (4.8)$$

Where, $F(t) = \int_0^t f(t)dt$ is the cumulative disrtibution function.

Failure Rate

$\emptyset(t)$ is failure rate function at time t, given that it survived until time t:

$$\emptyset(t) = \frac{f(t)}{R(t)} \qquad (4.9)$$

Relationship between Failure Density and Reliability

$$f(t) = \frac{-d}{dt} R(t) \qquad (4.10)$$

Relationship between h(t), f(t), F(t) and R(t),

$$h(t) = \frac{f(t)}{R(t)} = \frac{f(t)}{1 - F(t)} \qquad (4.11)$$

Remark: The failure rate h(t) is a measure of proneness (degree of getting subjected of failure) to failure as a function of age, t.

Relationship between MTBF/MTTF and Reliability:

$$\text{MTBF} = \text{MTTF} = \int_0^\infty R(t)dt \qquad (4.12)$$

Example 4–5. Trial data shows that 105 items failed during a test with a total operating time of 1 million hours. (For all items, that is, both failed and passed). Also, find the reliability of the product after 1000 hours, that is, $(t) = 1000$

Solution: The failure rate λ=No of failures/total operating time

$\lambda = \frac{105}{1000000} = 1.05 \times 10^{-4}$ per hour

Reliability at 1000 hours $= e^{-\lambda t}$

$$R(1000) = e^{-\left(1.05x10^4 x 1000\right)} = 0.9$$

Therefore the item has a **90%** chance of surviving for 1000 hours

Example 4–6. The data below shows operating time and breakdown time of a machine.
Operating Time is 20.2, 6.1, 24.4, 35.3 and 46.7
Down Time is 2.5, 7.1, 4.2 and 1.8

a) Determine the MTBF.
b) What is the system reliability for a mission time of 20 hours?

Solution:
Total operating time = 20.2 + 6.1 + 24.4 + 4.2 + 35.3 + 46.7
= 136.9 hours
The failure rate λ = No of failures (Down time iterations)/total operating time
λ = 4/136.9 = 0.02922
Therefore, θ = MTBF = 1/λ = 34.22 hours
Now let us calculate the system reliability
R = $e^{-\lambda t}$ where t = 20 hours,
Substituting the values R = $e^{-(0.02922)(20)}$
R = 55.74%

Maintainability:

Maintainability is the measure of the ability of a system/instrument to be retained or restored to a specified condition when maintenance is performed by qualified personnel using specified procedure and resources. Maintainability can be measured with Mean Time To Repair (MTTR), MTTR is average repair time and is given by

$$\text{MTTR} = \frac{\text{Total Maintenance Down Time}}{\text{Total Number of Maintenance Action.}} \quad (4.13)$$

Also, MTBMA is an acronym for Mean Time Between Maintenance Actions including preventive and corrective maintenance tasks.

Researchers can use Software's like ReliaSoft's, Weibull++, RGA and BlockSim-packages which are useful for illustration and calculations. These software's are readily available online. Simulation software helps an engineer to determine the sample size, test duration or expected number of failures in a test. To determine these variables, analytical methods need to make assumptions such as the distribution of model parameters. In case of small sample size the simulation method is more accurate than the analytical method because it does not require any assumptions. Generally, a manufacturer has to prove that at any given point of time that a certain product has met a goal of certain reliability with a specific confidence.

Procedure to Use Weibull++ Software for Reliability Test:

The procedure to use weibull++ software is illustrated in the following steps. Four examples are shown in following section.

(i) **Determining Units for Available Test Time**

If duration of available testing facility is known, then to achieve reliability goals it is required to determine the number of units to test for this amount of time with no failures.

Example: Given that the parametric binomial method is used to design a test to demonstrate a reliability of 90% at 100 hours with a 95% confidence with zero failure during the test. Determine the units for available test time.

Assume a Weibull distribution with a shape parameter $\beta = 1.5$.

$$R = e^{-(t/\eta)^\beta}$$

$$\eta = \frac{t}{(-\ln(R_{DEMO}))^{\frac{1}{\beta}}} = 448.3$$

$$R_{TEST} = e^{-\left(\frac{t_{TEST}}{\eta}\right)^{\beta}} = 96.6\%$$

Substitute the appropriate values into the cumulative binomial equation given here:

$$1 - CL = \sum_{i=0}^{f} \frac{n!}{i!\,(n-i)!} * \left(1 - e^{-\left(\frac{t_{TEST}}{\eta}\right)^{\beta}}\right)^{i} * \left(1 - e^{-\left(\frac{t_{TEST}}{\eta}\right)^{\beta}}\right)^{n-i}$$

$$n = 85.49 \text{ or } = 86.$$

(ii) **Determining Time for Available Units**

Consider a case where 20 units are to be tested. The value of the scale parameter, η, are already been determined in previous example. The binomial equation with the Weibull distribution for t_{TEST} can be solved since the values of n, CL, f, η and β are already calculated in previous example. This value is 126.43[7]

(iii) **Test to Demonstrate MTTF: (Parametric Binomial Method)**

To determine the number of units to test for 60 hours a test is to be designed to demonstrate a MTTF= 75 hours with a 95% confidence with zero failure during the test $f=0$. Assume a Weibull distribution with a shape parameter $\beta=1.5$.

From the following MTTF equation determine the value of the scale parameter η. $MTTF = \eta \cdot \Gamma\left(1 + \frac{1}{\beta}\right)$

Where:

$\Gamma(x)$ is the gamma function of x. This can be rearranged in terms of η:

$$\eta = \frac{MTTF}{\Gamma\left(1 + \frac{1}{\beta}\right)} = 83.1$$

Since MTTF and β have been specified η can be calculated as 83.1. The value of R_{TEST} is calculated as:

Basic Instrumentation

$$R_{TEST} = e^{-\left(\frac{t_{TEST}}{\eta}\right)^{\beta}} = 54.1\%$$

By solving the binomial equation for n, the value is calculated as 4.88 or n = 5 units (The fractional value must be rounded up to the next integer value).

(iv) **Test to Demonstrate MTTF: (Non-Parametric Binomial Method)**
For non-parametric demonstration test design the same binomial equation can be used.

The time value is not associated with non-parametric binomial method. Hence it can be assumed that the value of R_{TEST} is associated with the amount of time for which units were tested. The given non-parametric binomial equation is widely used in practice when the available test time is equal to the demonstration time:

$$1 - CL = \sum_{i=0}^{f}(n_i)(1 - R_{TEST})^i R_{TEST}^{n-i}$$

Where:
CL is the confidence level,
f is the number of failures,
n is the sample size,
R_{TEST} is the demonstrated reliability,
R_{TEST} can be simply written as R if reliability is not related to time.
Example:
Determine the required sample size with zero-failure using non-parametric binomial method to demonstrate a reliability of 80% at a 90% confidence level.

4.5.5 Designing for Higher Reliability

A number of techniques are available to enhance the system reliability. Some of the important techniques are[8]:

(i) **Safety Critical Design**: There is a relation between product reliability and safety. In case of a product performing a safety-critical role, then failure of a key component can have direct consequences. There are different approaches to minimizing or preventing the risk of catastrophic failure.

(ii) **Over-specification**: In the building and construction industry for product applications, "x5" factor is mandatory for safety factoring all material strength calculations. Example: to have a suspension bracket for a 10kg light fitting, the system should be designed to carry at least 50kg.

(iii) **Redundancy (parallel)**: Multiple identical machineries are used distinctly, any one of which would be proficient of supporting usual product function. Examples: Data processing systems, Protective systems for nuclear reactors, Satellite communication subsystems, interconnected power systems, Aircraft propulsion systems, temperature control systems for space vehicles, Ignition systems for rocket engines. For example, a passenger lift has 4 cables carrying the lift cabin, all share the weight. Any one cable would be proficient of carrying the full nearside lift load. A catastrophe of up to threecables will not risk the lift inhabitants. Flight regulates and tool systems in some aircraft adopt alike policy. Dual cabling in military systems advances survivability.

(iv) **Redundancy (standby)**: A back-up scheme is held in stand-in and arises into operation only when the main system fails, for sample stand-by generators in hospitals, and stand-in parachutes. Standby redundancy is often pronounced as a "belt & braces" method.

(v) **Fail-safe design**: Assumes an integrate risk of failure for which the rate of any of these three approaches would be excessively high. The merchandise or system is designed to drop into a safe state in the event of partial or total failure.
For example:

- The gas supply to a local central heating boiler is closed off in the nonappearance of a "healthy" signal from the water pump, flame sensor, water pressure sensor, or exhaust fan
- Toys can be designed to crack at predetermined weak points so as to leave no sharp forecasts that would injure a child
- Railway train brakes are released by vacuum and applied by applying air. If a brake pipe bursts, the admitted air automatically applies the train brakes

Design faults create revise costs. Faults detected during the component plan phase can be adjusted locally without serious effect on development times and costs[9]. However, with each following stage in development, the rework cost rises by a factor of ten as shown in Table 4.2.

Through testing at each step the manufacturer tries to keep cost of the product low, especially during prototype and preproduction phases.

4.5.6 Reliability and Cost

From user point of view, the most rational criteria for deciding best system/instrument is by its cost. The two main component of this cost are purchasing cost (capital cost) and maintenance cost throughout its useful life. The reliability and purchase cost of the instrument/system goes on increasing, because of better design, quality control production technique. Hence for higher reliability the maintenance cost comes down. The total sum of these two keeps going down up to a point and then starts increasing again after a long period of time due to wear out of systems/instruments[10].

TABLE 4.2

Correction Errors with Respect to Stage of Production

Stage of production	Rework Costs (correcting errors)
Component design and manufacture	x1
Prototyping and subassembly manufacture	x 10
Final product assembly	x 100
Volume product shipped to dealers	x 1000
Volume product sold to customers and in use	x 10,000

The cost of various method of achieving the reliability will vary according to:

- Components used
- Maintenance cost
- Associability of product for the maintenance
- Time and man power availability for design
- Constraints such as weight, volume, and so on

The initial cost (design and production) increases, but the operating cost decreases with the reliability and hence there exist value of reliability for which the cost is minimum.

4.5.6.1 Product Law of Reliability

The system/instrument comprises of many subsystems or many components. Failure of any one subsystem or components leads to failure of the total system. This is series reliability problem. This is best explained by taking closed connections of switches or relays in series. It is assumed that, the failure is dependent and one failure leads to failure of other items. In practice, it is possible that failure of one component may lead to failure of other components. Where, Rx, Ry... shows the reliability of individual units. Therefore, Rs=Rx*Ry. Hence if the reliability of an individual subsystem or component is known then, the total system reliability is the product of individual reliabilities of all items.

Example 4.7 Consider two subsystems with the reliability of 85% and 90% (i.e. 0.85 and 0.9) then total subsystem reliability is given as:

Solution: R_s = 0.9*0.85 =0.765 or 76.5%

4.5.7 Environmental and Other Physical Effects on Reliability

The environment, in which an instrument or a system operates, has a profound effect on its reliability. This is true whether the instrument is operating (active), switched off or stored. In fact, when switched off, the effect of environment may be more damaging because there is no self-generated heat to counteract the effect of moisture. Following are some major environmental factors that affect the reliability.

(i) **Temperature**: The most important parameter is temperature. When the temperature rises above the normal value the rate of chemical reaction increases. It also speeds up oxidation process in the presence of increased moisture. Higher temperature results in development of stress caused by expansion. At very low temperatures, material hardens and become brittle.

Components like electrolytic capacitors become ineffective because of freezing of an electrolyte. As the temperature goes up, the electrolyte starts drying and thus reduces capacitor value permanently. The temperature effect that proves to be most damaging is those of rapid temperature cycles. This may happen when the equipment is in an unpressurised aircraft.

(ii) **Humidity**: The percentage of humidity in atmosphere can vary from 3% to 96% RH.

The main effects of humidity are:
- Reduction in the value of insulation resistance, which leads to electrical breakdown
- Corrosion due to moisture forming an electrolyte between dissimilar metals
- Reduction in insulation due to fungi growth

(iii) **Mechanical vibrations and shock**: An instrument experiences some degree of mechanical shock and vibration when transported and handled. Portable instruments like mobile equipment suffer the most. These effects can weaken the supports, loosen connections and wires and may physically break components, which inevitably are the reasons of failures and less reliability in the long run.

(iv) **Variations in pressure**: Low pressure causes a decrease of electrical breakdown voltage between contacts using air insulation, which can reduce reliability of the system/instruments. For illustration two case studies of a reliability test performed by Delserro Engineering Solutions, Inc. (DES) [http://www.desolutions.com/blog] are described below:

4.5.7.1 Case Study 1: Medical Product Reliability Testing

A leading medical equipment manufacturer needed help to set reliability goals and perform product reliability testing on their new product. DES helped them identify reliability goals. After the goals were set, DES developed a custom product reliability test plan that defined how the tests would be performed and how the test results would be quantified into a field life. DES designed and built automated test setups and performed life cycle testing on different components of their product. The test results were reported and analyzed with reliability computer software to determine the product's field life or Mean Time to Failure (MTTF)[11].

4.5.7.2 Case Study 2: Temperature and Environmental Product Test

A manufacturer of consumer products was looking for a way to quantify their product's field life. Specifically, they were looking for a way to quantify the degradation of their product vs. time in the field. After discussing their project with DES, DES designed a test plan to expose their products to cyclic temperature and humidity. Severe use field conditions were defined and the conditions were accelerated during testing. Samples were periodically removed during the test and measurements were made. A statistical analysis was performed to demonstrate that the product would not degrade significantly over the customer's 15 year lifetime requirement.

4.5.7.3 Case Study 3: Reliability and Failure of Electrocardiogram (ECG) Machine

Following case study provides researcher basic understanding about performance characteristic under operating conditions. In the previous section reliability and failure of instruments and how it affects the performance are discussed. In this case study 12-lead electrocardiogram is taken and will discuss how diagnosis gets affected by few errors and how it can mislead to false results[11].

The 12-lead electrocardiogram (ECG) is extensively used tool for the diagnosis and identification of cardiac condition which mainly include myocardial infarction and ischemia[12].

The correct placement of each electrode plays vital role for correct patient diagnosis as shown in Figure 4.13 Any incorrect placement of leads mainly leads to a wrong diagnosis which may cause wrong ECG result. ECG waveform under normal condition for healthy person is shown in Figure 4.14. Standard placement of electrode is given in Table 4.3.

4.5.7.4 Error Due to Lead Reversals

Lead switch placement is a common mistake when ECGs are taken which mainly leads to false diagnoses of patients giving false results.

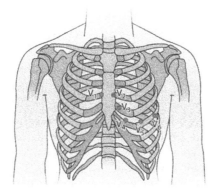

FIGURE 4.13
12-Lead Placement for ECG[12]

TABLE 4.3

Position of Leads in ECG

ELECTRODE	PLACEMENT
RL	Place anywhere above the ankle and below the torso
RA	Place anywhere between the shoulder and the elbow
LL	Place anywhere above the ankle and below the torso
LA	Place anywhere the shoulder and the elbow
V1	Place in the 4th intercostal space to the right of the sternum
V2	Place in the 4th intercostal space to the left of the sternum
V3	Place at midway between V2 and V4
V4	Place at 5th intercostal space at the mid-clavicular line
V5	Place at anterior axillary line at the same level as V4
V6	Place at mid-axillary line at the same level as V4 and V5

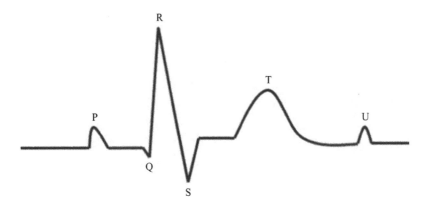

FIGURE 4.14
ECG Waveform under Normal Condition for Healthy Person

- Left-right arm inversions prompt a negative (-ve) complex in lead I with a negative (-ve) P wave in lead I. This is the essential reason for right axis deviation which occurs on the ECG waveform
- Arm-foot changes prompt a little or "far field" motion in lead II or III
- Chest lead inversions prompt wrong R wave progression in ECG waveform

So if any right axis or even a little change in a furthest point lead ought to be one reason enough to check lead position. Any disturbance or an error caused due to electrical interference and movement of leads is called as "Artifact".

As filter settings impact the interpretation of the ECG and furthermore decrease electrical interference of ECG machine. Provision for both low pass filter and high pass filter settings is made in the ECG machine. A high-pass filter is mainly used to decrease low frequency noise. It diminishes base line drift in the ECG waveform whereas a low-pass filter is mainly used to decrease high frequency noise (delivered by chest, extremity muscles where electrical interference is produced by the power grid). In monitor mode the high-pass filter which is principally set at 0.5–1.0 Hz and the low-pass filter on the 40 Hz frequency. This is the most attainable filter setting and it allows only narrow frequency signals to pass through these filters. This setting is helpful for rhythm monitoring in that noise can divert ST segment, which is not that important. In this mode, pacemaker spikes seem to be invisible which further gets filtered on.

Diagnostic mode in this the high-pass filter is set at 0.05 Hz frequency and the low-pass filter at 40Hz, 100 or 150 Hz frequency. This improves the diagnostic accuracy and reliability with minimum error in the ST segment. On the other side a base line drift is obtained more easily.

Interference of the monitored ECG is very common among operating theatre and medical care unit environments, that is, (ICU). Art factual signals, that corrupt the conventional cardiac signal, might arise due to few internal sources and few external sources. Many electrical devices employed in the clinical applications will introduce artifacts by numerous completely different mechanisms. All these artifacts could also be nonspecific or might tally serious arrhythmia. Easy measures, comparable to correct attention to basic principles of medical instrument measure, will eliminate some artifacts. In persistent cases, skilled facilitate could also be needed to spot the precise supply and minimize interference on the ECG. Technological advancements in ECG signal could also be helpful to discover and reduce artifacts. Ultimately, associating improved understanding of the artifacts generated by an instrument, and their distinctive characteristics, is very important to neglect misdiagnosis, misinterpretation and induced complication.

Reliability of an instrument plays key role in research. The selection of the instrument to produce measurement of variables must be reliable and valid which will lead to consistent and accurate foundation of the element in research. Selection of an appropriate data collection instrument is an important component of research finding because, a basic element of validity and reliability of the study depends on the chosen qualitative method (e.g., questionnaire) or quantitative method (e.g., survey). Data collection, analysis and formulation are the important aspects in the research. Following section describes types of data and the different methods of data collection with advantages, disadvantages and applications.

4.6 Data Collection

Data collection is the process of collecting and measuring information of selected variables in an established systematic manner. The systematically collected data enables researcher to answer relevant questions and evaluate outcomes. Data collection is important factor of research in all fields of study including physical and social sciences, humanities, Engineering and Technology and business. Data collection methods vary by discipline, the emphasis on ensuring accurate and honest collection is common[13]. The goal of data collection is to get quality evidence which will allow researcher to analyse formulation of convincing and credible answers to prove his/her hypothesis. Data collection begins after finalization of research problem, accurate data collection is necessary for maintaining the integrity of research. Inaccurate data causes consequences that includes the inability to answer research questions accurately and the inability to repeat and validate the research.

4.6.1 Types of Data

To decide the method of data collection a researcher should know types of data. A Researcher would have to decide which kind of data is required for research and then accordingly he will have to decide the method of data collection. To reduce likelihood of errors occurring selection of appropriate data collection instruments (existing, modified, or newly developed) and clearly delineated instructions for their correct use are needed. The data has been divided into two types: primary and secondary.

(i) **Primary data**:

The data which are collected fresh for the first time are called as primary data. This may include number of ways like interviews, focus groups, telephone surveys, and so on.

Primary data can be collected from large population and across wide geographical coverage through emails and posts.

4.6.1.1 Advantages of Primary Data

- The primary data is original and pertinent to the subject of the research to achieve high accuracy in research
- The researcher can get a realistic view about the topic under consideration from the primary data
- Primary data are highly reliable because these are collected by the researcher or concerned and reliable party

4.6.1.2 Disadvantages of Primary Data

- To collect primary data coverage is restricted in size and for larger coverage a more number of researchers are required
- The cost of the data collection will increase with increase in time and efforts of more people but the importance of the research may drop
- Primary data collection takes lot of time and efforts. The problem of the research can become serious or out dated before completing the process of data collection, analysis and report preparation. It may defeat the purpose of the research
- In collection of survey based data one can face design problems while preparing questionnaire. The prepared questionnaire must be simple to understand and respond
- Timely responses are not received from some respondents

Respondents may give socially acceptable, fake answers to suppress the realities.

In primary data collection incomplete questionnaire can give a negative impact on research.

Skilled personnel are required for data collection. Unskilled person in data collection may give inadequate data of the research

 (ii) **Secondary data:**

The data which has already been collected by someone else, analysed and statistically processed is called as secondary data. It has been collected by someone not related to the current research field but collected this data for some other motive and at different time in the past. If the researcher uses this data to make conclusions then this becomes secondary data for the researcher. Secondary data may be available in written, typed or in electronic forms. Sources of secondary information are available to the researcher for assembling data on an industry, market place and potential product applications. The researcher gains an initial insight into the research problem from secondary data. Secondary data can be of internal or external type. If information is acquired within the organization where research is being carried out the data is called as internal or in-house secondary data. If data is obtained from outside sources then data is called as external secondary data. Advantages and disadvantages of using secondary data are listed here.

4.6.1.3 Advantages of Secondary Data

- Secondary data is economical and faster to access
- It gives a way to use the work of the best intellectuals all over the globe
- It imparts a persuasion to the researcher about direction he/she should follow for the selected research topic
- Secondary data adds value to the research study by preserving time, effort, and money

4.6.1.4 Disadvantages of Secondary Data

- The data collected by someone else may not be as reliable and as accurate compared with primary data
- Data collected at one location may not be acceptable for the other location because of inconsistent environmental conditions
- As time passes the data becomes obsolete and antiquated
- Collected secondary data may pervert the results of the research
- To use secondary data a special care and permissions are necessary to amend or modify for use
- To use secondary data, care should be taken to avoid issues of copyright and authenticity

Considering the benefits and limitations of sources of data, demand of the research topic and time factor, both the data sources, that is, primary and secondary can be selected. These are used in combination to give proper coverage to the topic.

4.6.2 Sources of Data Collection

After knowing types of data, a defined data collection process is required to ensure that the data gathered are both systematic and accurate and that succeeding decisions based on arguments embodied in the findings are valid. The data collection process gives an idea about a baseline from which to measure and in certain cases an indication of what to improve. Classification of sources of data collection is represented in Figure 4.15.

Information gathered from primary sources is collected using first approach. To collect primary data there are three sources viz. Observation of research field, interviewing group of people and sending questionnaire to group coming under research area.

(i) **Observation**: A complex source of primary data collection is observation. As it forces researcher to do multitasking and to use a number of techniques continuously; including her/his five senses to collect data. The objective of observation method is to observe people in their natural ways as they follow their everyday lives. Observation methods can be preferred because it can overcome the issues like validity, bias in quantitative research methods. These methods are primarily useful when its subject can provide inaccurate information. Out of available methods for collecting primary data for the topic of study survey and observation methods have been found suitable. Observation can be participant or non-participant based on participation of researcher.

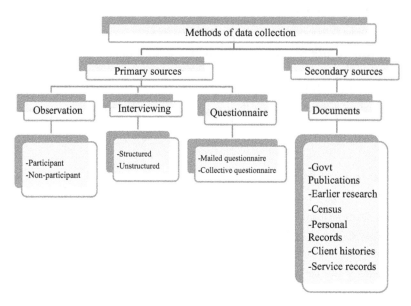

FIGURE 4.15
Sources of Data Collection

- **Participant**: The researcher observes actual situation carefully by putting himself in that situation. On the basis of relevance to research topic, his/her skills, knowledge and experience he collects the data from respondents by observing their natural behaviour. The results of participant observation completely depends on the talent and skills of the researcher. This method can be used only by experts and experienced researchers
- **Nonparticipant**: It is a data collection technique whereby the researcher does not take an active part in the situation under scrutiny only watches the subjects of his or her study, with their knowledge. The data obtained can be invalid as observed people may behave differently hence this approach is sometimes criticized

4.6.2.1 Advantages

- Subjective bias can be eliminated
- Data is unaffected by past behaviour or future intentions
- Natural behaviour of group is allowed to record

4.6.2.2 Disadvantages

- Gives limited amount of information
- Unseen factors can affect the observation task

(ii) **Interviewing**:

In this method of primary data collection the interviewer (researcher) personally communicates with the informants and asks required questions to them about the research topic. Data can be efficiently collected from the informants by cross examining them. Efficient and tactful interviewer is required to retrieve relevant and accurate data from the informants. Interviews can be structured or unstructured. As per the need of the study personal interview, depth interview or telephone interview can be conducted.

- **Structured**: A set of questions or a questionnaire related to the topic of research study is prepared by the researcher and questions have been asked according to that. It gives quantitative data and evaluates explicit data. This method is applicable to validate information
- **Unstructured**: Unstructured Interview is an interview in which the questions to be asked to the respondents are not prepared in an advance. This method is unrehearsed and very casual. It depends on conversation which tends to focus respondents personal qualities related to the work. Questions can be asked about skills, knowledge and strengths and should be answered formally similar to structured interview. It gives qualitative information and evaluates implicit information. This method is applicable to judge if the candidate is right person for job

4.6.2.3 Advantages

- This method allows the researcher to gather information from illiterate humankind
- As interviewer collects data personally there are nil chances of non-response
- As interviewer tactfully collects the data by cross examining the responders hence collected data is very reliable

4.6.2.4 Disadvantages

- It has chances of bias
- The informants may avoid answering few personal questions
- The process is time-consuming
- Requirements of money and manpower are very high

For successful implementation of interviewing, the researcher should take the following care to improve the accuracy of the data collected. The interviewer should be carefully selected. The interviewer and respondent must be attentive, and observant. They must possess the technical competence and necessary practical experience. Occasional field checks must be arranged to ensure that interviewers are not cheating or deviating from instructions given to them. Interviewer should be able to answer legitimate question(s) if asked by the respondent and must clear any doubt that the latter has. Efforts should be taken to create friendly atmosphere of trust and confidence, so that respondents can feel ease while talking and discussing with interviewer. The interviewer must ask questions intelligently and properly and must record the complete response.

(ii) **Questionnaire**:

A set of questions which has been prepared to ask and collect answers from respondents related to the topic of research is called a questionnaire. Individuals need to answer questions usually in printed or electronic form. Questionnaires are mailed questionnaire and collective questionnaire. To obtain statistically useful information about a given topic a series of questions have been asked to individuals. The feedback forms often prepared with blank spaces where answers can be written. Sets of such forms relating to research topic have been distributed among groups and the answers are collected.

Properly constructed and responsibly administered, questionnaires can become a vital instrument to make conclusions about specific groups or people or entire populations. The survey may become valueless if inappropriate questions, incorrect ordering of questions, incorrect scaling, or bad questionnaire format are used, as it may not accurately reflect the views and opinions of the participants. To validate questionnaire and making sure it is accurately recording the required information a pre-test among a smaller subset of target respondents can be suggested. Questionnaire helps in achieving four basic purposes:

- Collect the appropriate and required data
- Make data differentiable and biddable to analysis
- Reduce bias in constructing and interrogating questions
- Make questions engaging and interesting

Mailed questionnaire:

The respondents who are located at a long distance or across borders and other communication ways are difficult then mailed questionnaire is used for collection of data. The required thing is that the researcher should have the postal addresses of the respondents. The questionnaire can be posted or mailed to the respondents them, but in every case it will return to the researcher via mail. At their own convenience respondents can answer the questions. The detailed information can be collected for the research purpose as respondents cannot be biased by the researchers.

4.6.2.5 Advantages

- Large amounts of information can be collected from group of people in cost effective way
- Can be carried out by the researcher or by any number of people with limited affect to its validity and reliability
- Quick and easy quantification of the results can be done by either a researcher or through the use of software
- Scientific and objective analysis can be done effectively compared to other forms of research

4.6.2.5 Disadvantages

- The response rate is very less due to low literacy rate or lack of interest in the research topic

- It is criticized to be inadequate to recognize some forms of information –, that is, changes of emotions, behavior, feelings, and so on
- Gives only a limited amount of information without any explanation of it
- Truthfulness of a respondent is difficult to measure

Sample Questionnaire

Following example shows questionnaire for taking feedback of students about resource person called for expert lecture at college /university departments. To rate expert person called for lecture five scales are defined as strongly agree, agree, no opinion, disagree, and strongly disagree. For successful implementation of questionnaire the researcher should take the following care to improve the accuracy of the data collected.

The researcher must be clear about the various aspects of his/her research problem to be dealt with in the course of research project. A Researcher should re examine invariably and in case of need must revise the questionnaire. Pilot study must be carried out for pretesting the questionnaire. Questionnaire must have simple and straight forward directions for the respondents. A sample questionnaire is shown in Figure 4.16.

Secondary sources are equally important in research. It contains already available data. It includes government publications, earlier research, census, personal records, client histories, and service records.

4.7 Scaling

It is the process of generating the continuum, a continuous sequence of values, upon which the measured objects are placed. Scaling techniques are classified as comparative scaling techniques and noncomparative scaling techniques.

4.7.1 Comparative Scaling

It gives direct comparison of stimulus objects. Data is interpreted in relative terms and is measured on ordinal scale. This technique is non-numeric technique. These are the easy techniques to apply. The respondents have to choose between two brands of cold drink "A" and "B" even if there is very small difference in their liking of two brands. The main disadvantage is inability to generalize beyond the stimulus. Example, if someone wants to compare third brand of cold drink "C" with A and B, we have to conduct new study.

4.7.2 Noncomparative Scaling

In this method of scaling each object is scaled independently of the remaining. Example respondents are asked to give preference on 1 to 5 scales, 1: not preferred at all while 5: highly preferred. Respondents are asked to give score for cold drink brand A and B on 1 to 5 scales. To compare C, score can be collected for C on the same scale.

Student Feedback

1. The Guest Lecture helped me to assess my current skills:
   S.A ☐   A ☐   N.O ☐   D ☐   S.D ☐

2. The Guest Lecture has clear goal and meaningful content:
   S.A ☐   A ☐   N.O ☐   D ☐   S.D ☐

3. The Guest Lecture facilitator was knowledgeable and approachable:
   S.A ☐   A ☐   N.O ☐   D ☐   S.D ☐

4. The Guest Lecture was organised at appropriate time:
   S.A ☐   A ☐   N.O ☐   D ☐   S.D ☐

5. Time given for this Guest Lecture was
   1. Too Short ☐   2. Too long ☐   3. Right Length ☐

6. Please rate the Following:

	Excellent	Very Good	Good	Fair	Poor
i. Visuals	☐	☐	☐	☐	☐
ii. Acoustics	☐	☐	☐	☐	☐
iii. Presentation and PPTs	☐	☐	☐	☐	☐
iv. The overall program	☐	☐	☐	☐	☐

7. The Workshop/Seminar/Guest Lecture/ Industrial Visit taught me new skills and strategies:
   (# Choose from following according to you subject so that CO/PO get satisfied#)

	Excellent	Very Good	Good	Fair	Poor
a. Modern Tools Usage	☐	☐	☐	☐	☐
b. Engineering in social context	☐	☐	☐	☐	☐
c. Environment & Sustainability	☐	☐	☐	☐	☐
d. Ethics	☐	☐	☐	☐	☐
e. Individual & Team work	☐	☐	☐	☐	☐
f. Communication	☐	☐	☐	☐	☐
g. Lifelong Learning	☐	☐	☐	☐	☐
h. Project management & Finance	☐	☐	☐	☐	☐

8. What did you most appreciate/ enjoy/ think was best about the Guest Lecture?
   _____

Name: (S.A= Strongly Agree,
                                      A= Agree,
                                      N.O= No opinion,
                                      D= Disagree,
Sign: S.D=Strongly Disagree)

FIGURE 4.16
Sample Questionnaire

Basic Instrumentation

Q. How would you rate contents of Newschannel

Version 1:
The worst --X---------- The best

Version 2:
The worst --X---------- The best
  0   20   40   60   80   100

Version 3:
The worst --X---------- The best
            20   40   60   80   100
         Very poor   Neither poor   Very good

FIGURE 4.17
Linear Scaling

4.7.3 Linear Scaling

It is also called as continuous rating or graphic rating. It comes under non-comparative scaling technique. Various points are usually put along the line to form a continuum and it rather indicates his/her rating by marking on line which runs from one extreme to another having opposite values as shown in figure below. An example in Figure 4.17 represents survey of contents of news channels.

4.7.3.1 Limitations

A respondent can choose any position along the line which may increase the difficulty of analysis. The meaning of the terms like "very unsatisfied" and "very satisfied" may depend on respondent's frame of reference so that the statement might be challenged in terms of its equivalency.

4.8 Role of Digital Signal Processing in Data Collection and Noise Filtering

Data Analysts in various realms such as industrial, biomedical, genetics, geographical, astronomy, government and commercial observe the challenge of managing rapid growing volumes of data that are collected for research. Mining valuable information from large number of collected samples needs to adopt inventive tactics that efficiently process the data and utilize their structure. When data collection and analysis is discussed then the most important part is the methodology which is picked for this purpose[14]. If the methodology is properly selected then the data will be appropriately analyzed or sorted which leads researcher to end up with effective and logical conclusion. In research process it is likely to have the data which is noisy or redundant which is intrinsically created due to unwanted junk data or specifically for engineering researchers it is due to unwanted signals in the process.

While dealing with collected data it is challenging how to deal with noise suppression, pattern matching, random deletion in data stream for example in bioinformatics due to many junk DNA it is possible that two DNA sequence can code same gene or in speech or written text data analysis various adjectives can be deleted or added. It may occupy the wide frequency range and can affect current and voltage conditions. Characteristics of noise determine performance of system. In instrumentation and measurement system effect of noise plays a critical role as it affects the accuracy.

4.8.1 Sources of Noise

Operational reliability of system mostly depends on the environmental factor called as noise. Noise can be:

- Random in nature
- Repetitive
- Continuous
- Independent burst

To eliminate or reduce noise it is important to understand the sources of noise. Sources of noise can include internal sources, external sources, and locally produced noise.

(i) **Internal Noise Sources**

This type of noise is generated by components of electronics system. Following are some examples of noise generated by internal noise sources.

- White Noise: This type of noise affects equally all the frequencies. It has Gaussian amplitude distribution
- Shot Noise (Poisson noise): It is a type of electronic noise which can be modelled by a Poisson process. Shot noise originates from the discrete nature of electric charge. Shot noise also occurs in photon counting in optical devices, where shot noise is associated with the particle nature of light
- Thermal Noise: This is the electronic noise generated by the thermal agitation of the charge carriers (usually the electrons) inside an electrical conductor at equilibrium, which happens regardless of any applied voltage
- Flicker Noise: This is a type of electronic noise with a $1/f$ power spectral density. This noise increases with decrease in frequency so it is also called as $1/f$ noise
- Barkhausen Noise: If the system contains some magnetic sensors then due to finite size of domains of ferromagnetic materials the noise get introduced, that noise is called as Barkhausen noise
- Contact Noise: When there is breakdown of contacts between adjacent particles in current path the noise introduced is called as contact noise

(ii) **External Noise sources**

This type of noise is caused due to external factors or due to other electrical systems.

- Switching current and voltages: When high current load switches turns on and off they cause transients in power supply lines because of their inductance and capacitance
- Power Lines interference: When a high current is flowing through the line parallel to signal line it produces noise even if it is few distance away
- Sparking and radiation: Electromagnetic wave as a noise is produced whenever there is spark or arching. Unshielded lines and open conductors are also causes noise
- Environmental and atmospheric noise: Lightening discharge in thunderstorm causes spurious radio waves in the form of impulses that affects radio broadcasting
- Electrostatic discharge: This serious problem observed in areas where there is much dryer air, air conditioned buildings, and continental climates

(iii) **Local Noise Sources**

This type of noise is associated with circuit interconnections, board interfacing, transmission cables, and power supplies. This can be reduced by providing attention while designing the circuits.

4.8.2 DSP in Noise Filtering

To prove research hypothesis it is of prime importance to use noiseless data or avoid use of redundant data. To achieve noiseless data different signal processing techniques are used. Vast spectrum of digital signal processing which is now going beyond the limits of electronics field made it more popular among all the streams of engineering research and its expansions have vividly impacted modern society everywhere. DSP application spectrum includes but not limited to digital/Internet audio or video, digital recording, CD, DVD, and MP3 players, digital cameras, digital and cellular telephones, digital satellite and TV or wire and wireless networks. A medical instrument provides useful information for precise diagnoses with the help of digital electrocardiography (ECG) analyzers or digital x-rays and medical image systems that uses DSP. DSP is a powerful tool which scientists, engineers, and researchers can use to analyze and visualize data and perform research.

Following are the typical examples of research area where DSP can be used to process noisy data.

(i) **Geophysical data signal processing**

Geophysical signal analysis deals with the detection and processing of signals. Any varying signal conveys valuable information, to understand the information entrenched in such signals, it is necessary to "detect" and "extract data" from such signals. Geophysical signals are information bearing signals which carry data related to petroleum deposits beneath the surface and seismic data hence they are of immense important. Analysis of geophysical signals gives qualitative understanding of occurrence of a natural catastrophe such as earthquakes or volcanic eruptions.

(ii) **Structural health monitoring in civil engineering**

In many parts of the world incidents such as building and bridge collapse are on rise without little apparent warning. It has become of immense importance to develop methods to detect the damage/degradation of the civil structures as the number of such incidences has rapidly increased. To avoid impending disability or collapse buildings and critical infrastructure are monitored like a patient in a hospital, for signs of degradation. Sensors are used to monitor state of the health of the structures and technologies like human brains to analyze the abnormal situation [15].

(iii) **Medical Instrumentation**

Images are nothing but the signals with special characteristics. Parameters of image signals are measured over space, that is, distance and contain great information hence in medical instrumentation field this subgroup of digital signal processing gained importance. Magnetic Resonance Imaging uses magnetic field and radio waves to probe the interior of human body. The strength and frequency of fields are carefully adjusted so that atomic nuclei in a localized region of body resonate between quantum energy states because of this secondary radio waves are emitted. This emitted signal is captured by antenna placed near the body and it provides important information and it is in the form of image. This technique provides discrimination between different types of soft tissue. This MRI technic provides important information regarding blood flow through arteries. MRI totally depends on digital signal processing.

(iv) **Seismology**

In reflection seismology a sound pulse is sent into the ground. This pulse produces a single echo for each layer through which it is passed. While returning to the surface of earth it has to pass through all other boundary layers above where it is originated. This results into echo bouncing and it will produce echoes of echoes. The signal detected at the surface is very complex and difficult to analyze as it contains secondary echoes. Digital signal processing is widely used to isolate primary echo from secondary echo. The reflection seismic method is primary method for locating petroleum or mineral deposits. Seismic researcher commonly uses DSP for data collection, analysis and filtering. Fast Fourier Transform is used to differentiate between natural seismic signal and nuclear test explosions as they have different frequency spectrum.

(v) **Genomics**:

In recent times, DSP has gained huge popularity in data analysis of Genomics. In this field DSP analyzes the DNA sequence data, identifies the coding and non-coding regions and finds out abnormalities in coding region based on filtering. A Discrete Fourier Transform–based spectral approach can be used for this purpose and Digital Infinite Impulse Response Low Pass Filter with Butterworth approximation is used for enhanced data analysis and filtering.

(vi) **Digital Filtering**

It is a difficult task to represent a digitized noisy signal obtained from sensor output containing a useful low-frequency signal and noise that occupies all of the frequency range. After ADC, the noisy signal that is digitized with a number of samples can be improved using digital filtering. The signal of interest contains low frequency components and high frequency components above the interested frequency component considered as noise, this noise can be eliminated using a digital low pass filter. The signal that is a digitized noise signal is now a digitized clean signal, this signal can be applied to appropriate DSP algorithm for a concern application or it can be given to digital to analog converter and reconstruction filter.

(vii) **DSP in Electrocardiography**

Problem with ECG signal recording is interference of 60Hz/50Hz power line signal. Usually it can be minimized by proper grounding and using twisted pair cables but the appropriate method is to use digital notch filter. This digital notch filter removes the specific frequency while keeping all other needed information. By this enhanced ECG recording accurate diagnoses can be done. This digitized noise filter can also remove 60-Hz interferences in audio systems.

(viii) **Signal processing applications in Automotive**

Nowadays in automotive systems various types of communication protocols such as Controller Area Networks (CAN), Local Interconnect Network (LIN) and others are used. These controllers mainly use digital signal processing chips for signal processing purpose. The processing can be done with 8-bit microcontrollers (MCUs), DSPs and field programmable gate arrays (FPGAs). 8- and 16-bit MCUs are hardly used today because of their limited signal processing capabilities. Often to compromise costs system developers will choose a processor with performance that is just enough to perform to specific task. In some applications it may be sensible to build performance-based system rather than cost. Automotive researchers will use DSP algorithms for different infotainment challenges in automotive sector. Data from various nodes in automotive is collected and analyzed. This reference data can be used for further optimized operations.

The following two case studies will make the role of DSP in data collection and analysis to give correct prediction of result.

Case Study I: Structure of bird song

Classification of bird species according to specific category is now advanced application of DSP. Several methods are proposed and marked as successful for specifically labeling the species of single birds among thousand other noisy environments[16].

Lawrence Neal, Forrest Briggs, RavivRaich, and xiaoliZ.fern_school of EECS, Oregon State University, Corvallis have proposed a method of preprocessing and segmenting noisy field recordings of bird song to isolate each bird syllable from rest of the signal[17].

The song of bird consist of single vocalization syllables and such many syllables remains relatively constant therefor the method of species recognition is mainly based on classification at the level of syllables[18].

(i) **Data Collection**

The data required for the research to identify structure of bird song is collected from audio recordings of bird songs consisting of sequential bird call syllables separated by silence. Method of classification consist of characteristics of an audio signal such as Mel-frequency ceptral coefficients (MFC) (MFC is a representation of the short-term power spectrum of a sound, based on a linear cosine transform of a log power spectrum on a nonlinear mel scale of frequency) to label each syllable in the audio signal as a bird species.

The process of segmenting or extracting syllables from an audio signal is simple in ideal conditions. If the bird call is only source in a signal then increased audio energy will denote a syllable but in practical condition this assumption does not work .Along with bird song there can be many other unwanted sources of sound in a recording process such as wind, steam and other background noise. This unwanted signal or noise can degrade the required data. While recording vocalization from two or more individual birds may occur simultaneously, which is one of the complicated factors in data analysis because accurate segmentation is required for species classification.

There are two main challenges faced in noisy environment:
- Due to nonbird noises, false syllables are introduced as time-domain segmentation is based purely on audio energy
- Accuracy of classification degrades whenever multiple birds sing simultaneously at once because time-domain segmentation will group mixtures of syllables together

For this experiment author collected a four TB dataset of audio form recorded from automatic recorder placed on site (H.J.Andrews experimental forest).

Recording in this data set contains highly noisy data and simultaneous two or more bird syllables.

(ii) **Data Filtration**

First of all input signals are transformed into a spectrogram representation. After formation of spectrogram supervised classifier is applied to create a binary mask labelling each time-frequency unit as either bird sound or background. This process will ensure extraction of individual syllable of bird song even when syllable overlap in a time.

To transform the signal into time frequency spectrogram a short time FFT is applied with frame size of 512 samples and overlap of 256 samples between each subsequent frame. To normalize the level of environmental noise at each frequency a whitening filter is applied to the spectrogram.

(iii) **Results**

Frequency ranges below 1KHz contain little or no bird call so a band pass filter will remove such noisy data under 1kHz.

Case Study 2: Physiological sounds in cardiac cycle

The heart normally produces repeatable physiological sounds in one cardiac cycle. But in some pathologic conditions, such as with heart valve stenosis or a ventricular septal defect, blood flow causes turbulence which leads to the production of additional sounds which are called as murmurs. Murmur sounds are random in nature, whereas

Basic Instrumentation

underlying heart sounds are deterministic so these two signals have different spectrums.

(i) **Data Collection** Digital Subtraction Phonocardiography makes use of appropriate digital filters to separate the random murmur component spectrum of the phonocardiogram from the underlying deterministic heart sounds spectrum. The analytical data in the form of signals can be collected from:
- Different individuals with cardiac structural disease
- Recordings from normal individuals
- Individuals with innocent heart murmurs

(ii) **Data Filteration**

Data filteration, also called digital subtraction analysis of signals, is done with custom computer program called Murmur gram. The recorded sound from two adjacent cardiac cycles is subtracted to produce a difference signal using the Murmur gram.

(iii) **Results**

Advanced signal processing techniques can be used to separate heart sounds from murmurs.

4.9 Measuring Instruments and Tools in Engineering Discipline

The following table gives the details of various measuring instruments and tools used in engineering discipline which can be used by a researcher to perform experiments to prove his/her hypothesis. The following Table 4.4 gives a ready reference of various test and measuring instruments to the researchers who intend to work on interdisciplinary research problems.

4.10 Summary

This chapter gives a brief introduction to instrumentation schemes. Instrumentation schemes can be classified as simple instrumentation scheme or complex instrumentation scheme based on number of measuring and controlling the parameters.

After identifying the process variables based on rate of change of variables, the performance characteristics of an instrument are divided into two categories: Static characteristics and Dynamic characteristics. If quantities being measured are either slowly varying with time or constant are called as static characteristics and if quantities being measured are rapidly varying, then known as dynamic characteristics. Static characteristics includes Accuracy, Range, Sensitivity, Repeatability, Reproducibility, Drift, Hystersis, Resolution, Precision, Stability, and Tolerance, while Speed of Response, Measuring Lag, Fidelity, and Dynamic Error are dynamic characteristics.

TABLE 4.4
Various Test and Measuring Instruments Used in Engineering Discipline

Sr. No.	Name of Instrument	Parameter Measurement
ELECTRONICS AND COMMUNICATION ENGINEERING		
1.	Ammeter (Ampermeter)	Current
2.	Power Factor Meter	Power factor
3.	Multimeter	Resistance, Capacitance, voltage, current, and so on
4.	IC tester	To detect faulty ICs
5.	Distortion meter	It displays the amount of distortion added to the original signal by an electronic circuit.
6.	Electricity Meter/Energy Meter	The energy dissipated from the circuit
7.	Pulse Generator	To generate rectangular pulses.
8.	Function Generator	To generate electrical waveforms (sine, square, triangular, and sawtooth/ramp) over a wide range of frequencies
9.	LCR-Q Meter	Inductance, capacitance, and resistance and Q factor.
10.	Ohmmeter	Resistance
11.	Mega-ohmmeter	Electrical resistance of insulating components by testing their strength.
12.	Leakage Tester	The leakage across the plates of a capacitor
13.	Microwave Power Meter	The electrical power at microwave frequencies typically in the range 100 MHz to 40 GHz
14.	Oscilloscope (Cathod ray Oscilloscope)	Displays waveform of a signal, allows to measure frequency, time, voltage, offset
15.	Signal Analyzer	Amplitude and the modulation of a RF signal
16.	Spectrum Analyzer	To display frequency spectrum of signal.
17.	Sweep Generator	It generates constant-amplitude variable frequency sine wave signal that tests frequency response of it
18.	Voltmeter (DVM and VTVM)	The potential difference between two points in a circuit
19.	VU Meter Volume unit (VU) meter or standard volume indicator (SVI)	The signal level in audio equipment
20.	Wattmeter	Power
21.	Peak Programme Meter (PPM)	Level of an audio signal for AM radio broadcasting networks
22.	True Peak Programme Meter	The peak level of the waveform irrespective of its duration
23.	Quasi Peak Programme Meter (QPPM)	True level of peak if peak level exceeds up to certain milliseconds, that is, for shorter duration
24.	Sample Peak Programme Meter (SPPM)	Digital audio-which shows only peak sample values (up to 3 dB higher in amplitude). It may have either a "true" or a "quasi" integration characteristic.
25.	Galvanometer	To detect and indicate electric current
26.	Vibration Galvanometer	To detect AC currents in the frequency of natural resonance, as a null indicating instrument in AC bridge circuits and current comparators
27.	Thermo-Galvanometer	Small electric current
28.	Time-Domain Reflectometer (TDR)	To characterize and locate faults in metallic cables, To locate discontinuities in a connector, printed circuit board, or any other electrical path

(Continued)

TABLE 4.4 (Cont.)

Sr. No.	Name of Instrument	Parameter Measurement
29.	Time To Digital Converter (Abbreviated TDC)	Advice for recognizing events and providing a digital representation of the time they occurred
30.	Transistor Testers	To test electrical behaviour of transistors and solid-state diodes
31.	Bus Analyzer	To check, test, debug and validate designs of a hardware-based product, It is also used to examine communication interoperability between systems and between components, and clarifying hardware support concerns.
32.	Logic Analyzer	This instrument captures and displays multiple signals from a digital system. A logic analyzer can convert the captured data into timing diagrams, protocol decodes, state machine traces, assembly language, or can correlate assembly with source-level software. Logic Analyzers have advanced triggering capabilities.
33.	Network Analyzer	S-parameters, Y-parameters, Z-parameters, and h-parameters. Network analyzers are used to characterize two-port networks but they can be used on networks with an arbitrary number of ports.
34.	Vector Network Analyzer	This instrument enables the RF performance of microwave devices to be characterised in terms of network scattering parameters, or S parameters. It can measure both amplitude and phase properties It is also known as gain-phase meter or an automatic network analyzer.
35.	Scalar Network Analyzer (SNA)	This instrument can measure amplitude properties only. An SNA is functionally identical to a spectrum analyzer in combination with a tracking generator.
36.	Microwave Transition Analyzer (MTA) Or Large Signal Network Analyzer (LSNA),	Amplitude and phase of the fundamental component and its harmonics.
37.	Video Signal Generator	Predetermined video and/or television oscillation waveforms, and other signals as an output. It is used in the synchronization of television devices stimulates faults in, or aid in parametric measurements of, television and video systems. It includes horizontal and vertical sync pulses (in analog) or sync words (in digital). Generators of composite video signals (such as NTSC and PAL) will also include a color burst signal as part of the output.
38.	Sync Pulse Generator	Produces synchronization signals, with a high level of stability and accuracy. They also provide a master timing source for a video facility.
39.	Vector Signal Analyzer	It measures the magnitude and phase of the input signal at a single frequency within the IF bandwidth of the instrument. The primary use is to make in-channel measurements of known signals. Vector signal analyzers are useful in measuring and demodulating digitally modulated signals like W-CDMA, LTE, and WLAN. These measurements are used to determine the quality of modulation and can be used for design validation and compliance testing of electronic devices.

(*Continued*)

TABLE 4.4 (Cont.)

Sr. No.	Name of Instrument	Parameter Measurement
40.	VSWR Meter (Voltage Standing Wave Ratio Meter)	It measures the standing wave ratio in a transmission line. It can also be used to indicate the degree of mismatch between a transmission line and its load (usually a radio antenna), or evaluate the effectiveness of impedance matching efforts.
41.	Vectorscope	It is a special type of oscilloscope which is used in both audio and video, phase of the colors in color TV applications. Vectorscopes are quite similar in operation to oscilloscopes operated in X-Y mode; however those used in video applications have specialized graticules, and accept standard television or video signals as input.
42.	Videoscope Or Video Borescope	A flexible Videoscope or Video Borescopeis an advanced type of borescope, which has a very small CCD chip embedded into the tip of the scope. The video image is relayed from the distal tip and focusable lens assembly back to the display via internal wiring.
43.	Antenna analyzer	An antenna analyzer (also known as a noise bridge, RX bridge, SWR analyzer, or RF analyzer) is an instrument used to measure the input impedance of antenna systems in radio electronics applications.
44.	Receptacle tester	A receptacle tester or outlet tester is a instrument used to verify that an AC wall outlet is wired properly or not. The tester itself is small device containing a power plug and several indicator lights.
45.	Tail-pulse generator	This instrument simulates the broad range of pulses encountered in the nuclear field.
46.	Continuity tester	Identifies break in an electrical path between two points of a circuit.
47.	Marx generator	This instrument generates a high-voltage pulse from a low-voltage DC supply. Marx generators are used in high-energy physics experiments, as well as to simulate the effects of lightning on power-line gear and aviation equipment.
48.	Noise-figure meter	The noise figure of an amplifier, mixer, or similar device.
49.	Electrostatic fieldmeter	Noncontact electricity charge on an object and the electrostatic field of an object in volts.
50.	Equivalent Series Resistance Meter	The equivalent series resistance (ESR) of real capacitors.
51.	Distortion meter	It determines specific frequencies that cause distortion in electronic devices. The device is primarily used in audio-related equipment.
52.	Cable tester	Tests the strength and connectivity of a particular type of cable or other wired assemblies.
53.	Frequency counter	Counts the number of oscillations or pulses per second (Frequency) in a periodic electronic signal.
54.	Optical spectrometer	Properties of light over a specific portion of the electromagnetic spectrum, typically for spectroscopic analysis to identify materials.

(Continued)

Basic Instrumentation

TABLE 4.4 (Cont.)

Sr. No.	Name of Instrument	Parameter Measurement
55.	Jitterlyzer	Jitterlyzer combines functions of jitter measurement and protocol analysis, replacing a combination of BERT, jitter-generator, and oscilloscope
56.	Slotted line	Slotted lines are used for microwave measurements which consist of a movable probe to be inserted into a slot in a transmission line. SWR is measured by a slotted line.
57.	Fiberscope	It is a flexible bundle of optical fibre with an eyepiece on one end and a lens on the other which is used to examine and inspect small places such as the insides of machines, locks, human body, and so on
58.	Arbitrary Waveform Generator (AWG)	Generates electrical waveforms either repetitive or single-shot (once only). The resulting waveforms can be injected into a device under test and analyzed as they progress through it, confirming the proper operation of the device or pinpointing a fault in it.
59.	Bolometer	It is a thermal detector that changes its electrical resistance as a function of the radiant energy striking it.
60.	Pyrometer	This device uses the radiant energy on each side of a fixed wavelength of spectrum. The band is quite narrow and usually centered at 0.65 μm in the orange-red area of the visible spectrum
61.	Digital Pattern Generator	An electronic instrument that generates digital stimuli. The main purpose of a digital pattern generator is to stimulate the inputs of a digital electronic device.
62.	Doppler RADAR Device	It is specialized radar that uses the Doppler effect to produce velocity data about objects at a distance. This operation is done by bouncing a microwave signal off a desired target and analyzing how the object's motion has altered the frequency of the returned signal.
63.	Harmonic Analyser	Calculates the total harmonic content of a sine wave with some distortion, expressed as total harmonic distortion (THD).
64.	Nd : YAG Laser	Nd:YAG (neodymium-doped yttrium aluminium garnet; $Nd:Y_3Al_5O_{12}$) is a crystal that is used as a lasing medium for solid-state lasers.
65.	Vernier Caliper Scale	This device is a visual aid which allows measuring more precisely when reading a uniformly divided straight or circular measurement scale.
66.	Polarimeter	Measures the angle of rotation caused by passing polarized light through an optically active substance.
SOFTWARE		
67.	Keil software	The Keil is a complete software development environment for wide range of microcontroller devices.
68.	Cadence Tool.	To design the schematic and the layout of Analog, Digital and mixed signal circuits.

(Continued)

TABLE 4.4 (Cont.)

Sr. No.	Name of Instrument	Parameter Measurement
69.	MATLAB Software	It is a high-performance language that integrates computation, visualization and programming in an easy-to-use environment where problems and solutions are expressed in familiar mathematical notation.
70.	Microwind Software	Designing and simulating circuits at layout level.
71.	LabView Software (Laboratory Virtual Instrument Engineering Workbench)	A system-design platform and development environment for a visual programming language. LabVIEW programs/diagrams are termed virtual instruments (VIs). Each VI has three components: a block diagram, a front panel, and a connector panel.
72.	Scilab	Scilab is free, open source software that provides a powerful computing environment for engineering and scientific complex computation involving applications.
73.	Express PCB	It includes Express SCH Classic for drawing schematics and Express PCB Plus for circuit board layout.
74.	MultiSim	It is an industry-standard, best-in-class SPICE simulation environment that captures and simulates program, which is part of a circuit design programs, along with NI Ultiboard.
75.	P-SPICE (Personal Simulation Program with Integrated Circuit Emphasis)	It is a SPICE circuit simulator for simulation and verification of analog and mixed-signal circuits.
76.	PSIM	It is an Electronic circuit simulation software package, which is designed specifically for power electronics and motor drive simulations but can be used to simulate any electronic circuit. It uses nodal analysis and the trapezoidal rule integration as the basis of its simulation algorithm.
77.	ORCAD	It is electronic design automation (EDA) software tool, which is used to create electronic schematics and electronic prints for manufacturing printed circuit boards.
78.	High Frequency Structure Simulator (HFSS)	It is advanced antenna performance simulation software, which provides fast and accurate prediction of installed antenna patterns, near-fields, and antenna-to-antenna coupling on large platforms.
79.	CST microwave studio suite	It is a special tool for the 3D Electromagnetic simulation of high frequency components.
80.	Virtual instrument software architecture (VISA)	This instrument provides standard for configuring, programming, and troubleshooting instrumentation systems comprising GPIB, VXI, PXI, Serial, Ethernet, and/or USB interfaces.
81.	Network Simulator	In communication and computer network, network simulation is a technique where a software program models the behaviour of a network by calculating the interaction between the different network entities (routers, switches, nodes, access points, links, and so on). ns-1, ns-2, and ns-3 are discrete-event computer network simulators, primarily used in research and teaching.
82.	Packet Tracer	It is a cross-platform visual simulation tool which allows user to create different network topologies and imitate modern computer networks.

(Continued)

TABLE 4.4 (Cont.)

Sr. No.	Name of Instrument	Parameter Measurement
83.	WireShark	It is a free and open source packet analyser that is used for network troubleshooting, analysis, software, and communications protocol development.

BIOLOGICAL AND BIOMEDICAL ENGINEERING

84.	Pulse Oximeter	Pulse rate and oxygen content
85.	Spectrophotometers	Concentration of given chemical solution
86.	Blood Cell Counter	Counts no of blood cells(i.e., RBCs, WBCs, Platelets)
87.	Flame Photo Meter	The concentration of certain metal ions.
88.	Flow Meter	Flow rate or quantity of a liquid/gas passing through a pipe.
89.	Rate Meter	It is also known as frequency meter that shows the rate of pulses.
90.	Audiometer	Acuity of hearing
91.	Chloridimeter	It is used in determination of chlorides
92.	Ph Meter	Measures pH of unknown solution
93.	Electronic Thermometer	Measures temperature of object
94.	Bimetallic Strip Thermometer	Used in industries in temperature control devices.
95.	Flurometer	The intensity of fluorescence, used chiefly in biochemical analysis
96.	Electron Microscope	Uses a beam of electrons to create an image of the specimen.
97.	ECG Machine	To measure cardiac activities of living organisms.
98.	EEG Machine	To measure brain activities of living organisms.

INSTRUMENTATION ENGINEERING

99.	Gas Meter	The volume of fuel gases such as natural gas, liquefied petroleum gas.
100.	Planimeter (platometer)	To determine the area of an arbitrary two-dimensional shape.
101.	Eudiometer	In this device mixtures of gases can be made to react by an electric spark, used to measure changes in volume during chemical reactions.
102.	Distributed Control System (DCS)	It is a computerized control system for a particular plant usually with a large number of control loops, in which autonomous controllers are distributed throughout the system.
103.	Programmable Logic Controller (PLC)	It is an industrial computer control system that monitors the state of input devices continuously. It also makes decisions based upon given program to control the state of output devices.
104.	Data Acquisition Card	Data acquisition (DAQ) is the process of measuring an electrical or physical quantities such as voltage, current, temperature, pressure, or sound with a computer. A DAQ system consists of sensors, DAQ measurement hardware, and computerized programmable software embedded in it.

(Continued)

TABLE 4.4 (Cont.)

Sr. No.	Name of Instrument	Parameter Measurement
ELECTRICAL ENGINEERING		
105.	Tachometer	Speed of an engine
106.	Anemometer	To determine wind speed
107.	Barometer	The atmospheric pressure.
108.	Manometer	To determine pressure
109.	Dynamometer	Force, torque or power of different systems.
110.	Clamp Meter	It is an electrical test instrument that combines a basic digital multimeter with a current sensor embedded in it.
111.	Wheatstone Bridge	Electrical resistance
112.	Capacitance Meter	Capacitance of different capacitors specifically discrete type of capacitors.
113.	Ground Fault Protector	To open ungrounded conductors when high currents, especially those from line-to-ground are encountered in systems.
114.	Hygrometer	Humidity of the working environment.
115.	Shear Viscometer	Viscosity of a non-Newtonian fluid at several different shear rates.
116.	Thermocouple	Temperature
117.	Megger	Electrical leakage in wire.
118.	Varmeter	Reactive power of the circuit.
119.	Watt-Hour Meter	Electrical power passing through a circuit in a certain time.
120.	Vector Impedance Meter	Amplitude and phase angle of impedance (Z).
SOFTWARES		
121.	Mipower software	It is a highly interactive, user-friendly windows based Power System Analysis package which includes a set of modules for performing a wide range of power system design and analysis.
MECHANICAL ENGINEERING		
122.	Piezometer	Pressure of a liquid or gas. They are often placed in boreholes to monitor the pressure or depth of underground water.
123.	Tribometer	Measures tribological quantities, such as coefficient of friction, friction force, and wear volume between two surfaces in contact.
124.	Venturimeter	The rate of flow of a fluid flowing through a pipe.
125.	Pitot tube	It is a pressure measurement instrument which measures flow velocity of any liquid.
126.	Exhaust Gas Analyzer (exhaust CO analyzer)	To measure carbon monoxide from other gases in the exhaust, caused by an incorrect combustion.
127.	Salt Spray Chamber	The salt spray (or salt fog) is standard and most popular test for measurement of corrosion used to check corrosion resistance of materials and surface coatings.

(Continued)

TABLE 4.4 (Cont.)

Sr. No.	Name of Instrument	Parameter Measurement
128.	Surface Roughness Tester	This instrument is used to determine the surface texture or surface roughness of a material accurately. It shows the measured roughness depth (Rz) as well as the mean roughness value (Ra) in micrometers or microns (μm).
129.	Tensile Tester	Used for tension testing.

SOFTWARES

130.	Solid Edge Software	This software tool addresses all aspects of the product development process – 3D design, simulation, manufacturing, data management.
131.	Ansys V12 software	It supports product design and validation in a virtual environment that captures complex and coupled physical phenomenon.
132.	EdgeCAM and CADEM software	It is CAD-CAM software specifically used for 3D milling, mill turn, multi-axis machining, and 3D machining.
133.	AutoCad software	It is a computer-aided drafting software program that creates blueprints for buildings, bridges, and computer chips, and so on

POLYMER ENGINEERING

134.	Smoke Density Apparatus	It measures and observes the relative amount (density) of smoke produced by the burning (combustion) or decomposition of plastic andcables.
135.	Limiting Oxygen Index Tester	Determines the minimum oxygen in a flowing mixture of oxygen and nitrogen for flaming combustion.
136.	Conductivity Meter (EC meter)	Measures the electrical conductivity in a solution. It is commonly used in hydroponics, aquaculture, and freshwater systems to monitor the amount of nutrients, salts or impurities present in the water.
137.	UV Spectrophotometer	This instrument uses visible and ultraviolet light to study and analyze the chemical structure of substance. A spectrophotometer is a special type of spectrometer, which is used to measure relationship between the intensity of light and its wavelength.
138.	Fourier Transform Spectrophotometer	It is a measurement technique where spectra are collected based on measurements of the coherence of a radioactive source, using time-domain or space-domain measurements
139.	FTIR spectrometer (Fourier Transform infrared spectroscopy)	FTIR is a technique that provides an infrared spectrum of absorption or emission of a solid, liquid or gas. It simultaneously collects high-spectral-resolution data over a wide spectral range.

ENVIRONMENTAL ENGINEERING

140.	Turbiditimeter	The relative clarity of a fluid by measuring the amount of light scattered by particles suspended in a fluid sample.
141.	Nessler Tube	This device compares color and turbidity between solutions. The tubes are often used to carry out a series of calibration solutions of increasing concentrations.

(Continued)

TABLE 4.4 (Cont.)

Sr. No.	Name of Instrument	Parameter Measurement
142.	Water Testing Kit	It tests presence of bacteria, lead, pesticides, nitrites/nitrates, chlorine, hardness, and pH present in water.
143.	Electronic Balance	It measures the mass of a substance. It is used in experiments in which precise amount of measurements are required of every substance for desired results.
144.	BOD (biochemical oxygen demand) Test Device	It determines the effect of industrial waste and private household waste on the water quality in sewage treatment plants and their receiving waters.
145.	Chloroscope	This instrument checks presence of residual chlorine in drinking water.

CIVIL ENGINEERING

Sr. No.	Name of Instrument	Parameter Measurement
146.	Pycnometer	It is a kind of specific gravity bottle that determines the density of any liquid.
147.	Hydrometer	It is an instrument for measuring the density of liquids.
148.	Casagrande Device	This device determines the moisture content at which clay soils pass from plastic to liquid state.
149.	Jodhpur Permeameter	This instrument measures rapidly the electromagnetic permeability of samples of iron or steel with sufficient accuracy for many commercial purposes.
150.	Proctor Compaction Test Apparatus	It is a laboratory method of determining the optimal moisture content at which a given soil type will become most dense and achieve its maximum dry density.
151.	Optical Square	It is a square refracting block which refracts an incident beam at an angle of 90 degrees. It can be used with an autocollimator for measuring squareness of a workpiece.
152.	Prism Square	A hand optical square used by surveyors to lay off correct angles that are multiples of 90° or 45°.
153.	Line Ranger	An instrument for locating an intermediate point in line with two distant signals. It consists of two reflecting surfaces so arranged as to bring images of the two signals into coincidence when the instrument is in line with the signals
154.	Clinometer	The angle or elevation of slopes.
155.	Trough Compass	Used for marking the magnetic north line on the drawing sheet of the plane table
156.	Sprit Level	This device consists of a sealed glass tube partially filled with alcohol or other liquid has an air bubble whose position reveals whether a surface is perfectly levelled or not.
157.	Dumpy Level	It is a levelling optical instrument used to establish or verify points in the same horizontal plane. It is used in surveying and building with a vertical staff to measure height differences and to transfer, measure and set heights.
158.	Level Staff (Level rod)	Determines the difference in height between points or heights of points above a datum surface.

(*Continued*)

TABLE 4.4 (Cont.)

Sr. No.	Name of Instrument	Parameter Measurement
159.	Transit Theodolite	This device is used to measure horizontal and vertical angles. They can rotate along with their horizontal axis as well as vertical axis.
160.	Electronic Theodolite	This device provides highly flexible and accurate measurement of angles, alignments, grade work, and short range levelling.
161.	Tacheometer	This device is used specifically for rapid surveying, by which the horizontal and vertical positions of points on the earth's surface are determined without using a chain or tape or a separate levelling instrument.
162.	Le Chatelier's Test Device	It is used for determining the expansion of cement.
163.	Slup Test Device	The device is used for concrete that is too fluid (workable) to be measured because the concrete will not retain its shape when the cone is removed.
164.	Compression Testing Machine	This device maintains the compressive force or crush resistance of a material.
165.	Elongation Gauge	Elongation is increase in a sample's gauge length measured after a rupture or break divided by the sample's original gauge length.
166.	Flakiness Gauge	This device is used to evaluate if aggregate particles are to be considered flaky that means if their thickness is less than 0.6 of their nominal size.
167.	Ring And Ball Apparatus	This device determines softening point of bitumen.
168.	Los Angeles Test Device	It performs all procedure for testing coarse aggregates for resistance to abrasion.
169.	Penetrometer Test Set Up	It gives information on the geotechnical engineering properties of soil. The test uses a thick-walled sample tube that is driven into the ground at the bottom of a borehole by blows from a slide hammer.
170.	Digital Compression Testing Machine	To test concrete for its compressive strength.
171.	Atomic Absorption Spectrophotometer	The concentration of elements in a liquid sample based on energy absorbed from certain wavelengths of light.
172.	Rebar Corrosion Testing Facility	To measure rebar corrosion rate that helps for estimating the service life of concrete structures.
173.	Bond Strength Testing Facility	To determine the stress required to rupture a bond formed by an adhesive between two metal blocks.
174.	Goniometer	For precise measurement of angles, especially the angles between the faces of crystals.
175.	Graphometer	For angle measurements that consist of a semicircular limb divided into 180 degrees.
COMPUTER ENGINEERING		
176.	IOPM	This software defines a range of power management constants used in several in-kernel and user space APIs.
177.	IP load tester	It is a class of protocol analyzers tha tobserves the practical evaluation of router performance.

(Continued)

TABLE 4.4 (Cont.)

Sr. No.	Name of Instrument	Parameter Measurement
178.	Fedora 16 (formerly Fedora Core)	Fedora is a Unix-like operating system based on the Linux kernel and GNU programs (a Linux distribution).
179.	JDK	The Java Development Kit (JDK) is a software development environment used for developing Java applications and applets.
180.	NCTUNS	NCTUNS is a Linux-based network simulator/emulator that has a great deal of features such as the possibility to execute real-world applications without modifications, and provision to model a wide range of network devices using real TCP/IP network stack.
181.	Oracle	The Oracle database is an object-relational database management system.
182.	Eclipse	It is an integrated development environment (IDE) used in computer programming, and is the most widely used Java IDE.
183.	SQL	As per ANSI (American National Standards Institute), it is the standard language for relational database management systems.
184.	GFI (Ground Fault Interupter)	It is a Network Security Scanner and Monitor
185.	Packet Broker	It is used for bandwidth analysis and bandwidth calculations
186.	Microsoft Network Monitor	It is used for monitoring the network
187.	Nagios	It is used in the analysis of networks (network, server, and log monitoring) and database.
188.	Open NMS (Network Management System)	It is used for network management
189.	KVM (Kernel Based Virtual Machine)	It is used to run multiple virtual machines at the same time
190.	Delta cloud	It is used for managing heterogeneous clouds.
191.	Cloud9 IDE (Integrated Development Environment)	It is used to give different features to work together in team.
192.	Codenvy	Codenvy's team workspaces allow developers and stakeholders to collaborate on premergecode without installing any software.
193.	Eclipse PHP (Hypertext preprocessor) Developer Tools	It is used to support PHP IDE Projects.
194.	Net Beans	It is used for desktop and mobile web development.
195.	Eclipse ADT (Android Development Toolkit)	It is used to develop different versions of Android.
196.	JBoss	It provides support to open source application server.
197.	SDK and AVD Manager	It is an Android Virtual Device Manger used to create an Android application.
198.	DDMS (Dalvik Debug Monitor Server)	A debugging tool used in the Android platform (port forwarding, on-device screen capture, on-device thread, and heap monitoring)
199.	LogCat	It is a command-line tool that dumps a log of system messages, including stack traces when the device throws an error and messages written from app with the Log class.
200.	Draw 9-Patch	It used to design graphics in Android OS.

(*Continued*)

TABLE 4.4 (Cont.)

Sr. No.	Name of Instrument	Parameter Measurement
201.	FlowUp	It is used to monitor the overall performance of android and iOS mobile app and get in-depth insights on various key performance metrics.
202.	XAMPP: Cross-Platform (X), Apache (A), MariaDB (M), PHP (P) and Perl (P)	It is an open source cross-platform for a Web server solution stack.
203.	WAMP Server (Web Applications, Apache2, PHP and MySQL)	It is a local host server for Web development.
204.	Wordpress	It is a Web development tool.
205.	Filezilla	It is a cross-platform FTP application (for client and server).
206.	Zenmap	It is the official Nmap Security Scanner GUI. It used to discover hosts and services on a computer network, thus building a "map" of the network.
207.	Adobe Dreamweaver	Web design and development tool that enables visualization of Web content while coding.
208.	Firebug	It is used for live debugging, editing, and monitoring of any website's CSS, HTML, DOM, XHR, and JavaScript.
209.	Selenium	It is a suite of tools used to automate a Web browser across different platforms.
210.	OpenVAS (Vulnerability Assessment System)	It is used in the security world for vulnerability scanning purposes.
211.	Nikto	It is a web server scanner that tests web servers for dangerous files/CGIs and outdated server software.
CHEMICHAL ENGINEERING		
212.	Gas Chromatography	Used in analytical chemistry for separating and analyzing compounds that can be vaporized without decomposition.
213.	High Pressure Liquid Chromatography	Used in analytical chemistry to separate, identify, and quantify each component in a mixture.
214.	Cyclic Voltometry	Measures the current developed in an electrochemical cell.
215.	Fluroscence Spectrophotometer	It is a electromagnetic spectroscopy that analyses fluorescence from given sample.
216.	Biophotometer	It is used for quick and routine measurements of the optical density of samples at predefined wavelengths. A Xenon flash lamp is used as light source.
217.	Chemi Doc MP Imaging System	It is used for imaging and analyzing gels and western blots.
218.	Multiporator	Used for the transfection of eukaryotic cells and bacteria in human, animal cells, plants, bacteria, and yeast.
219.	Gel Doc	Used in molecular biology laboratories for the imaging and documentation of nucleic acid and protein suspended within poly acrylamide or agarose gels.
220.	Thermocycler (Thermocycler, PCR machine or DNA amplifier)	Used to amplify segments of DNA via the polymerase chain reaction (PCR).
221.	Centrifuge	To separate fluids of different densities.
222.	Vortex Mixer (vortexer)	To mix small vials of liquid.
223.	Probe Sonicator	High power, programmable system ideal for nanotechnology, cell disruption, and homogenization applications to agitate particles in a sample.

(Continued)

TABLE 4.4 (Cont.)

Sr. No.	Name of Instrument	Parameter Measurement
224.	Phrase Contrast Fluorescent Microscope	Converts phase shifts in light passing through a transparent specimen to brightness changes in the image. Phase shifts themselves are invisible but become visible when shown as brightness variations.
225.	Gel Electrophoresis	It is a used for separation and analysis of macromolecules (DNA, RNA, and proteins) and their fragments, based on their size and charge.
226.	Katharometer	The thermal conductivity detector (TCD) used in gas chromatography.
Metallurgy Engineering		
227.	Inverted Metallurgical Microscope	For inspection of grain size and the state of the metals.
228.	Image Analyser Software: Digimizer	It allows precise manual measurements and automatic object detection with measurements of object characteristics.

Reliability of an instrument has also been discussed in the chapter. It is a property of any measure, tool, test, or of a whole experiment. It's an estimation of how much random error might be in the values around the true value. The reliability characteristics are classified as Repairable Systems and Nonrepairable Systems. Parameters playing vital roles include Failure Rate, MTTF (Mean Time to Failure), Time to First Failure, MRL (Mean Residual or remaining Life), Availability, Failure Rate and Repair, MTBF (Mean Time Between Failures), and MRL (Mean Residual or remaining Life).

To prove a hypothesis, the researcher performs experiments. Experiments can be performed in laboratory, directly on the field or a researcher can observe the natural changes in variables being measured. Hence, the experimental setup can be laboratory-based, field-based, or a natural experimental setup.

The data collection considerations and choice of appropriate data source is also presented. Data collection, analysis and formulation are the important aspects in the research. Data collection begins after finalization of research problem, accurate data collection is necessary for maintaining the integrity of research. Data is classified as primary data and secondary data. Observation, interviewing, questionnaire are primary data sources and documents like government publications, earlier research, the census, personal records, client histories, and service records are some examples of secondary data sources. Selection of an appropriate data collection instrument is an important component of research finding because, a basic element of validity and reliability of the study depends on the collected data.

Scaling is the process of generating the continuum, a continuous sequence of values, upon which the measured objects are placed. Linear scaling also called as continuous rating or graphic rating is most commonly used scaling technique. Various points are usually put along the line to form a continuum and rating can be given by marking on line which runs from one extreme to another.

To prove research hypothesis it is of prime importance to use noiseless data or avoid use of redundant data. A brief summary of different noise sources is presented. To achieve noiseless data different signal processing techniques are in use. DSP application spectrum includes but not limited to digital/Internet audio or video, digital recording, CD, DVD, and MP3 players, digital cameras, digital and cellular telephones, digital satellite and TV or wire and wireless networks. Without DSP, scientists, engineers, and researchers would have no powerful tools to analyze and visualize data and perform their research.

List of measuring instruments and tools used in engineering discipline to prove a hypothesis is also presented at the end of the chapter.

Further Reading

1. D. V. S. Murty, "Transducers and Instrumentation", Second edition, New Delhi: PHI Learning Pvt. Ltd., 2008.
2. A. K. Sawhney, "Electrical and Electronic Measurement Measurements and Instrumentation", Fourth Edition, New Delhi: Dhanpat Rai & Sons Educational and Publishers, 2011.
3. E. O. Doebelin, "Measurement Systems", 6th edition, New Delhi: Tata Mcgraw Hill, 2012.
4. Joseph J. Carr, "Elements of Electronic Instrumentation and Measurement", Third Edition, New Delhi: Person, 2003.
5. B. C. Nakra, K. K. Choudhari, "Instrumentation Measurements and Analysis", New Delhi: Tata mcgraw Hill Education, 1985.
6. http://www.davehitt.com/facts/helena.html
7. http://reliawiki.org/index.php/Reliability_Test_Design
8. Andrew Taylor, "bsc MA FRSA – Art and Engineering in Product Design.
9. H. Choi, S. Choi, H. Cha, "Structural Health Monitoring System Based on Strain Gauge Enabled Wireless Sensor Node", Proceedings of the Fifth International Conference on Networked Sensing Systems, Kanazawa, Japan, June 17–19, pp. 211–214, 2008.
10. "Introduction to Reliability Engineering, e-Learning Course", CERN.
11. Joseph J. Carr, John M. Brown, "Introduction to Biomedical Equipment Technology", Fourth Edition, Person Education, 2002.
12. Andrew R. Houghton, Alun Roebuck, "Pocket ECGs for Nurses", CRC Press, 2015.
13. C. R. Kothari, G. Garg, "Research Methodology: Methods and Techniques", 4th edition, New Delhi: New Age International Publications, 2017.
14. "'Digital Signal Processing'-Fundamentals and Applications: Li TAN", 2nd edition, Devry University, Decatur, Georgia: Academic Press, Elsevier, 2013.
15. K. Maenaka, "MEMS Inertial Sensors and Their Applications", Proceedings of the Fifth International Conference on Networked Sensing Systems, Kanazawa, Japan, June 17–19, pp. 71–73, 2008.
16. Andrew J. Hansen, William C. Mccomb, Robyn Vega, Martin G. Raphael, Matthew Hunter, "Bird Habitat Relationships in Natural and Managed Forests in the West Cascades of Oregon", Ecological Applications, vol. 5, pp. 555–569, 1995.
17. Lawrence Neal, Forrest Briggs, Raviv Raich, Xiaoli Z. Fern, "Time-Frequency Segmentation of Bird Song in Noisy Acoustic Environments", 2011 IEEE International Conference on Acoustics, Speech and Signal Processing (ICASSP).
18. Frederick F. Gilbert, Rochelle Allwine, Spring Bird Communities in the Oregon Cascade Range. In Leonard F. Ruggiero, Keith B. Aubry, Andrew B. Carey, Mark H. Hufftech. (Eds.), "Wildlife and Vegetation of Unmanaged Douglas-Fir Forests", tech., Portland, OR: U.S. Department of Agriculture, Forest Service, Pacific Northwest Research Station, 1991 pp. 145–158.

5
Applied Statistics

Varsha K. Harpale and Vinayak K. Bairagi

CONTENTS

- 5.1 Introduction 162
- 5.2 Regression Analysis 163
  - 5.2.1 Regression Model 163
  - 5.2.2 Simple Linear Regression Analysis 164
  - 5.2.3 Linear Ordinary Least Square (OLS) Regression 166
  - 5.2.4 Multiple Linear Regression 167
- 5.3 Parameter Estimation 169
  - 5.3.1 Estimating the Parameters 170
  - 5.3.2 Point Estimate 170
  - 5.3.3 Method of Moments 171
  - 5.3.4 Method of Maximum Likelihood Estimation 171
  - 5.3.5 Sampling Distribution of the Mean (SDM) 171
  - 5.3.6 Performance Parameter for Estimation 172
- 5.4 Inferential Statistics 174
  - 5.4.1 Chi-Square Test 175
  - 5.4.2 T-Test Analysis 176
- 5.5 Univariate, Bivariate, and Multivariate Data Analysis 178
  - 5.5.1 Univariate Analysis 178
  - 5.5.2 Bi-Variate Analysis 181
  - 5.5.3 Multivariate Analysis (MVA) 182
  - 5.5.4 Univariate Analysis of Variance (ANOVA) 182
  - 5.5.5 Multivariate Analysis of Variance (MANOVA) 183
- 5.6 Principal Component Analysis 184
  - 5.6.1 Procedure of Computing Principal Components Is as Detailed Below ... 184
- 5.7 State Vector Machines 187
  - 5.7.1 Principle of Support Vector Machine 187
- 5.8 Uncertainty Analysis 190
  - 5.8.1 Uncertainty Analysis Using Probability Theory 190
  - 5.8.2 Uncertainty Analysis of Linear and Non-Linear Systems 191
  - 5.8.3 Uncertainty of Measurement 191
  - 5.8.4 Uncertainty Analysis 192
  - 5.8.5 Uncertainty Calculation Example 193
  - 5.8.6 Software Tool for Uncertainty Analysis 194
- 5.9 Modelling and Prediction of Performance 195
  - 5.9.1 Concept of Mathematical Modelling 195
  - 5.9.2 Classification of Models 196

5.9.3   Stages of Modelling 196
5.9.4   Process of Modelling 197
5.9.5   Advantages and Disadvantage of Mathematical Modelling 198
5.9.6   Example of Mathematical Modelling: Integrated Solar Water Heater. . . . 198
5.9.7   Performance Measurement, Analysis, and Performance Curve 200
5.10 Multi-Scale Modelling. 201
   5.10.1   Approaches to Analyse Multiscale Structures 201
   5.10.2   Linear and Nonlinear Analysis of System 202
   5.10.3   Comparison of Linear and Nonlinear System 204
5.11 Sensitivity Analysis. 204
5.12 Statistica Software Tool. 204
5.13 Summary 205
Further Reading. 205

God Grant Me The Serenity To Accept The Things I Cannot Change, The Courage To Change The Things I Can And The Wisdom To Know The Difference.
—Dr. Reinhold Niebuhr

Learning objectives of this chapter are to

- Demonstrate descriptive and inferential statistics
- Apply linear regression analysis for estimating the relation between dependent and independent parameters
- Analyse the theory of estimation for promoting inferential statistics
- Evaluate the impact of sensitivity and uncertainty analysis in the mathematical model
- Introduce statistical tools such as principal component analysis and support vector machine
- Develop an ability to construct linear, nonlinear, and generalized mathematical models

This chapter will enable the researcher to:

- Understand the importance of regression models, estimation theory, statistical tools, and design of the mathematical model
- Select and formulate the statistical model for estimated parameters
- Formulate research problem using mathematical and statistical methodologies and solve using relevant structures
- Apply the properties of parametric, semiparametric, and nonparametric testing procedures and hypothesis testing for validating estimated models

5.1 Introduction

This chapter highlights the role of statistics in research for collecting and analyzing small scale or large data. Statistics is a branch of mathematics used to demonstrate research findings, support hypotheses, give credibility to research methodology and draw conclusions. It helps the researcher in understanding how to plan, conduct, and evaluate the research. The sections in this chapter will detail the selection of statistical methodology, includes a selection of suitable measures or parameters, a selection of evaluation methodology and analyzing data for drawing reliable conclusions by defining appropriate system models.

5.2 Regression Analysis

Regression concept was first introduced by F. Galton in the nineteenth century for biological research. The regression method was further extended by integrating it with various methodologies such as nonparametric regression, Bayesian-based regression, logistic regression, and polynomial regression. The regression involves estimation of dependent variables on the basis of controlled variables, and their relation using statistical modeling. The relation is observed by means of curves, images, and graphs/plots.

Thus, regression analysis is an estimation tool for anticipating conditional average values of dependent variables with reference to the constant value of the independent variable.

Sometimes in rare scenarios, the focus is on the location of the conditional distribution of the dependent variable with respect to the independent variables. Regression analysis is a most preferred method of defining the interrelation of variables in machine learning–based prediction or forecasting models.

Regression may be parametric regression, in which the regression function is defined in terms of a finite number of unknown parameters that are estimated from the data or nonparametric regression. The regression function lies in a specified set of functions which may use infinite-dimensional space. The performance of this method depends on the selection of data generation method, regression approach, assumptions, and types of variables. It is proven to be the most important tool for modelling and analysis of data.

For example, the rate of heart attack and affected age group of patients is one of the regression problems.

The benefits of regression analysis are as follows

- It estimates the **effective relationships** between the dependent variable and independent variable
- It derives the **effect** of multiple independent variables on a dependent variable

5.2.1 Regression Model

To estimate these independent variables a linear model is define called regression model, it contains following parameters and variables:

- The undefined scalar or vector parameters also called **unknown parameters (β)**
- The autonomous or explanatory or **independent variable, "X"**
- The reliant or **response variable, Y**

In regression model response variable Y is function of X and β as shown in equation (5.1),

$$Y \approx f(X, \beta) \qquad (5.1)$$

The approximation or estimation is given as **E(Y|X) = f (X, β)**. Proper selection of regression function "f" optimizes the performance of the regression analysis. Let β is a vector of the length "m" and consider information of dependent variable Y is a user-defined value then regression analysis can be performed in various condition as stated here.

Case 1: If data points of the form (Y, X) are "N" and it is observed that N < m then regression analysis is underdetermined due to the short length of data to calculate β.

Case 2: If function f is linear and it is observed that N=m, then Y ≈ f (X, β) and can produce an exact solution with linear values of X.

Case 3: If N>m then the system observed is overdetermined even though it provides enough information with a best fit value of β. This regression analysis is similar to the method of least squares where the distance between actual value and the predicted value of Y is minimized.

Let a regression model contain three unknown parameters, β_a, β_b, and β_c with 10 observations and independent variables X_1, X_2, and X_3. Due to lack of enough information, this regression model could not estimate unique values of β_a, β_b, and β_c. So calculate an average value and standard deviation of corresponding dependent variable Y. If regression of three different value of X is performed then it estimates three separate unknown parameters β.

In linear regression, if N>m and X^TX is invertible then the measurement errors are normally distributed and (N−m) shows an excess of information used for a statistical prediction about unknown parameters. This excess of information is referred as the degree of freedom of the regression.

5.2.2 Simple Linear Regression Analysis

The simple regression analysis is method in which single regressor is used to derive response of the system or to estimate the value of dependent variable. It is defined by equation (5.2) given below

$$Y = \beta_0 + \beta_1 X + \varepsilon \qquad (5.2)$$

Where β_0 is intercept and β_1 and slope of the behaviour

Assumptions of the linear regression model are as follows:

- For the prediction, the sample represents the population
- The error $'\varepsilon'$ is a random variable with a mean of zero conditional on the explanatory variables
- The autonomous variables are measured with no error thus $'\varepsilon = 0'$
- The autonomous variables are linearly independent

- The variance-covariance matrix of the errors is diagonal and each nonzero element is the variance of the error
- The variance of the error is constant across observations

The modified simple regression model is designed after applying all assumption
Sometimes spatial trends and spatial autocorrelation in the independent and dependent variables can violate statistical assumptions of regression.

Examples on Regression Analysis:

A randomly selected dataset of five students who appeared for mathematics test and scored as given below

- Find linear regression equation, which predicts statistic performance,
- Evaluate grading of the students with 80 marks in aptitude test
- Develop best fitted or estimated regression model for the system with calculation shown in Table 5.1

The regression function is linear and represented as $\left(\widehat{Y} = \beta_0 + \beta_1 X\right)$.

It is necessary to calculate values of b_0 and b_1 using formulas given below

$$\beta_1 = \sum \frac{[(Xi - \overline{X})(Yi - \overline{Y})]}{[(Xi - \overline{X})^2]} = \frac{460}{730} = 0.63$$

$$\beta_0 = \overline{Y} - \beta_1 * \overline{X} = 77 - (0.63) * (78) = 27.84$$

Thus, the calculated regression model is $Y = 27.84 + 0.63X$. This regression equation will describe the data by line of best fit using scatter plot as shown in Figure 5.1 drawn with an Excel tool where the equation of estimation is $Y = 25.76 + 0.654X$. This states correlation between marks and appeared students but optimized best-fitted line is not observed for this dataset but regression model can be estimated with a minimum error by approximating fixed intercept.

TABLE 5.1

Calculation of Linear Regression Model

Subject No.	No. of Students Xi	Measured Yi	$(Xi - \overline{X})$	$(Yi - \overline{Y})$	$(Xi - \overline{X})^2$	$(Yi - \overline{Y})^2$	$(Xi - \overline{X})(Yi - \overline{Y})$	Calculate: Predicted $\widehat{Y} = \beta_0 + \beta_1 X$	ERROR= $(Yi - \widehat{Y})$
1	60	70	−18	−7	324	49	126	65.64	4.16
2	70	65	−8	−12	64	144	96	71.94	−6.94
3	80	70	2	−7	4	49	−14	78.24	−8.24
4	85	95	7	18	49	324	126	81.39	13.07
5	95	85	17	8	289	64	126	87.69	−2.69
Sum	390	385			730	630	460		
Mean	78	77							

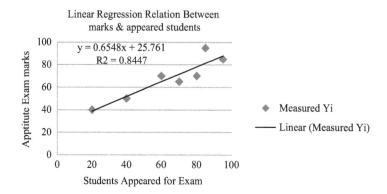

FIGURE 5.1
Line of Best Fit for Linear Regression Relation between Marks and Appeared Students

5.2.3 Linear Ordinary Least Square (OLS) Regression

Linear ordinary least squares (OLS) regression is score-based simple regression analysis and it uses the score of independent variable "X" to calculate the score of the dependent variable "Y". If these scores are linear then linear regression is preferable and can be applied as shown here:

- Linear straight line equation is used to determine regression line
- Determine the prediction score from this regression equation
- Evaluate the distribution of scores by calculating Standard Error of the Estimate(Sxy)

Data values "X" and "Y" are utilized to derive the regression equation using Pearson's rule. If the relation between the variables is significantly linear then selection of appropriate linear regression is important to estimate values of "Y" by the equation $\left(\widehat{Y} = \beta_0 + \beta_1 X\right)$. Where $\widehat{Y}$, a predicted score and β_1 is a slope of regression line with Y intercept β_0.

In OLS regression slope of the regression line is given by

$$\beta_1 = \frac{n\sum \overline{X} * \overline{Y} - \sum \overline{X} * \sum \overline{Y}}{n\sum \overline{X}^2 - (\sum \overline{X})^2} \tag{5.3}$$

Where $\overline{X}$ & $\overline{Y}$ are mean of the variables X and Y, respectively.
β_0 intercept is calculated by

$$\beta_0 = \overline{Y} - \beta_1 * \overline{X} \tag{5.4}$$

5.2.4 Multiple Linear Regression

As discussed earlier regression is a tool used to predict the performance of the system by estimating parameters and deriving the relationship between independent and dependent variables. In multiple regression models, there will be single dependent and multiple independent variables.

Let "x" series are for independent variables also called regressors and "y" is a dependent or response variable. In multiple regression, multiple regressors affect output response variable. If there are "k" numbers of regressors, then the model is represented with the linear equation given here:

$$y_i = \beta_0 + \beta_1 x_{i\,1} + \beta_2 x_{i\,2} + \beta_3 x_{i\,3} \ldots \ldots \ldots + \beta_{k-1} x_{ik-1} + \varepsilon_i \qquad (5.5)$$

Where $\beta_0, \beta_1, \beta_2, \beta_{k-1}$ are unknown coefficients that need to be estimated to define the model. ε_i is an error observed in the response and it is assumed to be following normal distribution, that is, $\varepsilon_i \sim (0,\ \sigma^2)$.

Estimation of these parameters defines regression model $Y = \beta X + \varepsilon$ of the system. The most commonly least square method is used for estimation of parameters and characterizes by the Squared Sum (SS) and Mean Squared Value (MS).

The estimated model is defined as

$$\widehat{y}_i = \widehat{\beta}_0 + \widehat{\beta}_1 x_{i\,1} + \widehat{\beta}_2 x_{i\,2} + \widehat{\beta}_3 x_{i\,3} \ldots \ldots \ldots + \widehat{\beta}_{k-1} x_{ik-1} \qquad (5.6)$$

Where estimated $\widehat{\beta}$ is calculated by

$$\widehat{\beta} = (X'X)^{-1} \times X'Y \qquad (5.7)$$

This model follows statistical properties as

- $E(\widehat{\beta}) = \beta$, the estimated parameter
- $V(\widehat{\beta}) = \sigma^2 (X'X)^{-1}$, the variance

The squared sum (SS) values for "n" observations of each of the regressor are calculated as

$$SS_T = \sum_{i=1}^{n}(y_i - \overline{y})^2 = \sum y_i^2 - n\overline{y}^2 \text{ with Df} = n - 1 \qquad (5.8)$$

$$SS_{Residual} = \sum_{i=1}^{n}(y_i - \widehat{y}_i)^2 \text{ with Df} = n - k \qquad (5.8)$$

$$SS_{Regression} = SS_T - SS_{Residual} = \widehat{\beta}'X'Y - n\overline{y}^2 \text{ with Df} = k - 1 \qquad (5.9)$$

The ideal model should have maximum $SS_{Regression}$ and minimum $SS_{Residual}$.

For Example:

Consider the following dataset containing two regressors and one response variable and define a model for this response and validate the model.

X1	X2	Y
1	8	6
4	2	8
9	-8	1
11	-10	0
3	6	5
8	-6	3
5	0	2
10	-12	-4
2	4	10
7	-2	-3
6	-4	5

Step 1: Define estimated model f1 or the system

$$Y = \widehat{\beta}_0 + \widehat{\beta}_1 x_1 + \widehat{\beta}_2 x_2 + \grave{o}$$

$$Y = \begin{bmatrix} 6 \\ 8 \\ 1 \\ 0 \\ 5 \\ 3 \\ 2 \\ -4 \\ 10 \\ -3 \\ 5 \end{bmatrix} \quad X = [1, x1, x2] = \begin{bmatrix} 1 & 1 & 8 \\ 1 & 4 & 2 \\ 1 & 9 & -8 \\ 1 & 11 & -10 \\ 1 & 3 & 6 \\ 1 & 8 & -6 \\ 1 & 5 & 0 \\ 1 & 10 & -12 \\ 1 & 2 & 4 \\ 1 & 7 & -2 \\ 1 & 6 & -4 \end{bmatrix} \quad \beta = \begin{bmatrix} \beta_0 \\ \beta_1 \\ \beta_2 \end{bmatrix}$$

Let estimate $\widehat{\beta} = (X'X)^{-1} \times X'Y = \begin{bmatrix} 14 \\ -2 \\ -1/2 \end{bmatrix}, \beta_0 = 14, \beta_1 = -2, \beta_2 = -1/2$

Thus, the fitted equation is $\widehat{Y} = 14 - 2x_1 - 1/2 x_2$

Step 2: Validating this fitted equation using Analysis of Variance (ANOVA)

a) Calculate SS_T, $SS_{Residual}$ and $SS_{Regression}$

X1	X2	Y	$\hat{Y} = 14 - 2x_1 - 1/2x_2$	$(Y - \hat{Y})^2$	$(\sum y_i^2) - n\bar{y}^2$	SS_T	$SS_{Residual}$	$SS_{Regression}$
1	8	6	8	4	9			
4	2	8	5	9	25			
9	−8	1	0	1	4			
11	−10	0	−3	9	9			
3	6	5	5	0	4			
8	−6	3	1	4	0	190	68	122
5	0	2	4	4	1			
10	−12	−4	0	16	49			
2	4	10	8	4	49			
7	−2	−3	1	16	36			
6	−4	5	4	1	4			

b) Calculate ANOVA table:

Source of Variation	Df	SS	MS= SS/Df	F score= MSReg./MSRes.
Regression	2	122	61	7.7
Residual	8	68	8.5	
Total	10	190		

c) State hypothesis using confidence level $\alpha = 0.05$

$H_0: \beta_1 = \beta_2$

$H_1: \beta_j \neq 0$ for at least one value of j

By F distribution table for confidence level $\alpha = 0.05$ and Df = 10, $F_{critical} = 4.46$

d) From the statistics, it is clear that $F_{estimated} > F_{critical}$ so reject H_0 hypothesis and the fitted equation can represent multiple linear regression model as $y = 14 - 2x_1 - 1/2x_2$

Thus linear regression estimates the coefficients of the linear equation to determine the relation between a response variable and explanatory variables. The regression analysis is of various types based on the distribution of the dependent variable. If "Y" is continuous then use linear regression model, if dichotomous use logistic regression, if the variable is Poisson or multinominal use log-linear regression model and if data is time-based with conditional cases then use Cox regression model for prediction of "Y".

5.3 Parameter Estimation

Statistical inference is an estimation theory used to estimate values of parameters such as mean, deviation, variance or correlation coefficients from randomly empirical data. Thus unknown parameters are estimated using measurement techniques [1]. The statistical inference facilitates estimating behavior of large data called population from sample data with a calculated degree of uncertainty. The well-known methods of statistical inference are an estimation and Null Hypothesis Test of Significance (NHTS). Hogg and Tanis (2001) explains, estimation is the process of obtaining numerical values of parameters or their range based on sample observation of the specific distribution. Outcome values are estimated and method used is called estimator. Estimators are random variables having their own probability distribution. For example, the process uses sample mean to estimate population mean (μ) and population proportion (p). Similarly, sample standard deviation is used to estimate population standard deviation (σ).

Properties of estimators include:

- If an individual parameter can estimate sufficiently the estimator is good
- Good consistency
- It may or may not be unbiased
- A minimum variance unbiased estimator is most sufficient estimator

For example, radars are used to find the range of the target by transmitting high strength signal and analyzing received echoes of transmitted pulses. Due to electrical noise, their actual values are randomly distributed, thus estimation of the transit time is a need of a model.

5.3.1 Estimating the Parameters

To find interrelation of variables and its performance, parametric estimation plays a very important role. Most commonly estimated parameters are:

- If the variance is known estimated parameter is amplitude <Aj>
- The noise variance based parameter, second moment <Aj, Ak>
- SNR (Signal to Noise Ratio)
- PSD (Power Spectral Density in watts/Hz)

The methods of parametric estimation are as discussed here.

5.3.2 Point Estimate

The statistical method used to estimate population from the single-valued sample is called "a point estimate" by Rohatgi [2]. A point estimate can be used for a population mean, a population variance, a population proportion (or) any other characteristic of the population.

Methods of point estimate most commonly used are method of moments, maximum likelihood estimation, minimum chi-square, least squares and minimum variance

Some desirable properties of point estimators include:

- Bias (η): The error observed in actual value and the estimated value of the parameter is a bias and given by $(E[\widehat{p_\eta}] - p)$

Applied Statistics 171

- Consistency: If the quantity of data increases the estimator is said to be consistent only in a case where it estimates $\hat{p}$ to cover true value of p
- Efficient Estimator: The efficiency of the estimator is defined by the variance of the estimators and lower the variance or correlation higher the efficiency of the estimator and is given by, $Var(\hat{p}_{n1})/Var(\hat{p}_{n2})$

The estimation method is based on finding most predicted parameter and can be measured by the difference between the observed sample covariance matrix.

5.3.3 Method of Moments

The method of moments uses the distribution function of a random variable to predict the expected value of parameters. It is based on comparing sample moments values with actual values of moments and solving this comparison could estimates the values of parameters. Let $E(X^k)$ is the kth(theoretical) moment of the distribution, and let is $M_k = \frac{1}{n}\sum_{i=1}^{n} X_i^k$ the kth sample moment for k=1,2...... Comparing these k moments, It estimates values of the parameters by solving the resulting $E(X^k) = M_k$. This estimator is consistent than any other biased estimator has less computational complexity.

5.3.4 Method of Maximum Likelihood Estimation

If a random sample $(x_1, x_2, x_3 \ldots x_n)$ of size n is selected from a population with Probability Density Function (pdf), f (xi,θ) for estimation then the likelihood function of the observed sample usually calculated from joint density function is L=L(θ)

For continuous random variable $L(\theta) = \prod_{i=1}^{n} f(xi, \theta)$, and

For discrete random variable $L(\theta) = \prod_{i=1}^{n} p(xi, \theta)$

Where p(xi, θ) is a probability mass function, L signifies relative likelihood for a set of samples and it is a function of the actual value of the parameter. Thus the method of maximum likelihood estimates the parameters with a maximum value of likelihood function L. It also follows the same properties as point estimates, that is, bias (low), efficiency (minimum variance), and consistency.

5.3.5 Sampling Distribution of the Mean (SDM)

A hypothetical distribution derives variability of $\overline{X}$ called Sampling Distribution of Mean (SDM) [3]. As in a point estimate, $\bar{x}$ gives reflections of parameter μ, but not the precision of the estimate. The probability of hypothetical outcomes is estimated from a hypothetical Probability Distribution Function (PDF).

The SDM divulge that:

- $\bar{x}$ is an unbiased estimate of μ;
- As the sample is huge and normalized then the SDM will tend to be normal (Gaussian)

- The predicted values of $\bar{x}$ are used to finds clusters around μ these statistics is the Standard Error of the Mean (SEM) and is given by:

$$\text{SEM} = \frac{\sigma}{\sqrt{n}} \quad (5.10)$$

Where σ is a standard deviation of the related external source of the size n.

For example, Let σ = 10 and n= 1, 4, 16 then by formula SEM is observed as 10, 5, 2.5 respectively. Thus for quadruple "n", the SEM observed to be half, this is called as square root law which states the relation between precision and the square root of the sample size [4].

5.3.6 Performance Parameter for Estimation

(i) **Interval estimate**.

A probable range of parameter can be estimated to within which its true value lies, it's called as interval estimate. For example $X_1 < X < X_2$ is an interval estimate of the population mean μ. The interval estimate is defined with a range of parameters and confidence level generally in a percentage, and it is associated with estimated ranges of a population parameter [1].

(ii) **Confidence Intervals (CI)** is a range of parameters within which unknown values are observed. In statistics, the confidence interval is used to define precision and uncertainty associated with various sampling methods. A confidence interval is characterized by confidence level, related statistics, and margin of error. Confidence intervals are preferred to point estimates because confidence intervals indicate the precision of the estimate and the uncertainty of the estimate.

The confidence level is the probability of finding confidence interval that includes the true population parameter. Let all possible samples are considered to estimate population parameters and confidence interval calculated for each one of the sample then some of the CI could include true population parameters. This is represented using confidence level such as 95% confidence level means that 95% of the intervals contain the true population parameter and so on.

The margin of error is the range of values above and below the sample statistic in confidence interval. For example, if an election survey is conducted in the newspaper then the reports states that 30% of the vote will be received by the independent candidate.

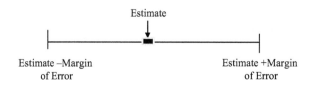

FIGURE 5.2
Confidence – Interval Range

The survey had a 5% margin of error and a confidence level of 95%, means surveyor is 95% confident that the independent candidate will receive between 25% and 35% of the vote.

As shown in Figure 5.2, the range of the estimate is calculated from (Estimate $\mp$ Margin of Error) and referred to as a confidence limit. The factor α is level of significance and $(1-\alpha)$ is called confidence coefficient which shows lack of confidence. This $(1-\alpha)*100\%$ represents the confidence level of a confidence interval and its value is depending on the precision of the estimate [5]. For example: Let there be n = 10 observations and a select sample of $\hat{x} = 29.0$ with SEM = 4.30, then calculate CI for μ with following formulas:

1) $SEM = \frac{\sigma}{\sqrt{n}}$
2) *Calculate z* quantile of $(1 - \alpha) = (z_{1-\alpha/2})$
3) Calculate $CI = \hat{p} \pm (z_{1-\alpha/2})(SEM)$

Sr. No.	Confidence level (CI for μ)	z quantile	$CI = \hat{x} \pm (z_{1-\alpha/2})(SEM)$	Margin of Error
1	90% α=0.10	$z_{1-0.10/2} = z_{0.95} = 1.64$	29.0 ± (1.64)(4.30) = 29.0 ± 7.1 = (21.9, 36.1)	±7.1
2	95% α=0.05	$z_{1-0.05/2} = z_{0.975} = 1.96$	29.0 ± (1.96)(4.30) = 29.0 ± 8.4 = (20.6, 37.4).	±8.4
3	99% α=0.01	$z_{1-0.01/2} = z_{0.99} = 2.58$	29.0 ± (2.58)(4.30) = 29.0 ± 11.1 = (17.9, 40.1).	±11.1

Thus the confidence level increases as margin of the error increases. Also, if variation in the population increases, the margin of the error increases, whereas if sample size increases, the margin of error decreases.

Steps involved in Estimation of a proportion:

Step 1. Review the research problem for identification of the parameters involved in the model. Select sample value that can address a binomial proportion (*p*).

Step 2. Point estimate. Calculate the sample proportion ($\hat{p}$) as the point estimate of the parameter.

Step 3. Confidence interval. Calculate confidence interval by $\hat{p} \pm (z_{1-\alpha/2})(SEP)$ of normal distribution and the standard error of the proportion $SEP = \sqrt{\frac{\hat{p}\hat{q}}{n}}$. Where $\hat{p}$ is point estimate and $\hat{q}$ is $(1-\hat{p})$ with sample size "n".

Step 4. Interpret the results. Estimate proportion variable relation and its confidence level with population parameter.

Illustration

In a survey of 2673 people it was observed that 170 have risk factor X. Calculate population prevalence of the risk factor with 95% confidence.

Step 1. Estimate parameter p.

Step 2. *Point estimate proportion* $\hat{p} = \frac{170}{2673} = 0.06360 = 6.36\%$.

Step 3.

    a) Apply "npqrule" for approximation and calculate $n\hat{p}\hat{q} = 2673(0.0636)(1 - 0.636) = 159$

    b) Calculate $SEP = \sqrt{\frac{\hat{p}\hat{q}}{n}} = \sqrt{\frac{(0.0636)(1-0.0636)}{2673}} = 0.00472$.

    c) Calculate z quantile for 95% of confidence level $z_{1-\alpha/2} = 1.96$

    d) Calculate Confidence Interval for proportion $p : \hat{p} \pm (z_{1-\alpha/2})\ (SEP) = 0.636 \pm 1.96 * 0.00472 = (0.0543, 0.0729) = (5.4\%, 7.3\%)$

Step 4. The prevalence in the sample was 6.4%. The prevalence in the population is between 5.4% and 7.3% with 95% confidence [4].

5.4 Inferential Statistics

Inferential statistics is a group of methods which draws a conclusion about the population from sample information using hypothesis testing. The sample is a randomly drawn subset of the population used to estimate population characteristics accurately.

Definition of inferential statistics

- It is "a method of reaching conclusions about unmeasurable populations by using sample evidence and probability"
- Vogt [6] define it as "Using probability and information about samples to draw conclusions ('inferences') about a population or about how likely it is that a result could have been obtained by chance."

A simple hypothesis test is used to draw the conclusion about the population such as chi-square test, T-test, and ANOVA, and so on. Hypothesis testing is used to estimate the population parameters or their relations. It is performed with the NULL hypothesis, H_0, and ALTERNATIVE hypothesis, H_1.

Procedure for Hypothesis Testing:

    Step 1: State Null hypothesis H_0 and alternative hypothesis H_1

    Step 2: Choose confidence or significance level for estimation, α

    Step 3: Calculate degree of freedom, Df

    Step 4: Find critical or standard distribution for defined confidence level from standard distribution table

    Step 5: Calculate test statistics and compare with its critical value for drawing a conclusion about the population

    Step 6: State the conclusion

5.4.1 Chi-Square Test

A test or squared distribution used in statistics to check whether sample exactly estimates the population or there is a specific relation between dependent and independent variables or to identify similarities in categorical variables.

The test is performed with chi-square coefficient, χ^2 and is given as:

$$\chi^2 = \frac{\sum (O_i - E_i)^2}{E_i} \qquad (5.11)$$

where O_i, is an actual value, E_i, is expected value and χ^2 is sample distribution with specific degree of freedom. The degree of freedom is the number of independent values or quantities that can be assigned to a statistical distribution. For chi-square test it is calculated from a number of columns, c and rows, r with formula Df=(c−1)x(r−1). PDF or chi-square distribution is as shown in Figure 5.3, which is used to make a decision of hypothesis rejection.

For example, a set of education levels are presented in a dataset of 415 persons for the male and female category so estimate the relation between education level and gender.

Category	HSC	Graduates	Post Graduates	PhD
Male	75	60	40	41
Female	45	51	55	48

Solution: First complete row and column wise analysis: (O_i)

	HSC	Graduates	Post Graduates	PhD	Total row wise
Male	75	60	40	41	216
Female	45	51	55	48	199
Total column wise	120	111	95	89	415

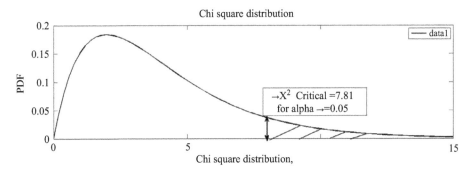

FIGURE 5.3
Chi-Square Distribution

Follow the steps of hypothesis testing:

Step 1: State hypothesis H_0 and H_1

H_0: Null hypothesis $P_M = P_F$

H_1: Alternative hypothesis $P_M \ne P_F$

Step 2: Select confidence level about 95% i.e. $\alpha = 0.05$

Step 3: Calculate Df = (c−1) × (r−1) = (4−1) × (2−1) = 3

Step 4: Find critical value from chi-square distribution table or contingency table for $\alpha = 0.05$ and df = 3. So $\chi^2 critical = 7.81$

Step 5: Calculate test statistics:

- Calculate estimated values of each observed value by the formula

$$E = \frac{Row\ total \times Column\ Total}{Sample\ Size} \quad (5.12)$$

So estimated values are as (E_i).

	HSC	Graduates	Post Graduates	PhD	Total row wise
Male	62.45	57.77	49.44	46.32	216
Female	57.54	53.22	45.55	42.67	199
Total column wise	120	111	95	89	415

- Now calculate χ^2 by the formula

$$\chi^2 = \frac{\sum(O_i - E_i)^2}{E_i} \quad (5.13)$$

Chi-square distribution for an individual is given as

	HSC	Graduates	Post Graduates	PhD	Total
Male	2.52	0.086	1.80	0.61	10.46
Female	2.73	0.092	1.96	0.66	

Step 6: State Conclusion:

As estimated $\chi^2 > \chi^2 critical$ and $p < 0.05$, reject the null hypothesis, which means there will be no relation between gender and education level, that is, they are independent.

5.4.2 T-Test Analysis

T-test analysis is most commonly used if the standard deviation of the population is unknown. Thus standard deviation can be estimated using T-test. It follows same steps of hypothesis testing with T-statistics as given here.

a) For single dependent variable

$$t = \frac{\bar{x}_D}{S_D/\sqrt{n}} \quad (5.14)$$

where $\bar{x}_D$ is a difference of mean and S_D is standard deviation difference with n number of sample size, which is calculated as

$$S_D = \frac{\sqrt{\sum x^2 - \frac{(\sum x)^2}{n}}}{n-1} \quad (5.15)$$

b) For two or more dependent variable

$$t = \frac{(\bar{x}_1 - \bar{x}_2)(\bar{\mu}_1 - \bar{\mu}_2)}{\sqrt{\frac{S_p^2}{n_1} + \frac{S_p^2}{n_2}}} \quad (5.16)$$

if μ_1 and μ_2 estimated means are given otherwise ignore the term as $(\bar{\mu}_1 - \bar{\mu}_2) = 1$ where $\bar{x}_1$ and $\bar{x}_2$ are mean of two independent variables and S_p^2 is given as

$$S_p^2 = \frac{SS_1 + SS_2}{Df_1 + Df_2} \quad (5.17)$$

SS is variation standard deviation as per the degree of freedom Df and calculated as $SS = Std^2 (Df)$ with Df as (n−1).

For example:

In a school, Class A and Class B scored marks as mentioned in the table. State their performance analysis.

Class	No. of Students	Average Score	Standard Deviation
Class A	25	70	15
Class B	20	74	25

Solution:

Step 1: Define hypothesis for the given data

H_0: $\mu_{ClassA} = \mu_{ClassB}$; performance of the Class A and Class B is same

H_1: $\mu_{ClassA} \neq \mu_{ClassB}$; performance of the classes is different.

Step 2: Consider confidence level 95% thus $\alpha = 0.05$

Step 3: Calculate degree of freedom

$$Df = (n_1 - 1) + (n_2 - 1) = (25 - 1) + (20 - 1) = 43$$
$$Df1 = (n_1 - 1) = 25 - 1 = 24,$$
$$Df2 = (n_2 - 1) = 20 - 1 = 19$$

Step 4: As per distribution table for $\alpha = 0.05$ and $Df = 43$ critical values are $\pm 2.5\%$ around the distribution, that is, ± 2.0167 (refer standard distribution table for T-test)

Thus if $-2.0167 < t$ estimated $< +2.0167$ then reject hypothesis H_0

Step5: Calculate test statistics: for two independent data of Class A and Class B

a) Calculate S_p:

$$S_p^2 = \frac{SS_1 + SS_2}{Df_1 + Df_2} = \frac{5400 + 11875}{24 + 19} = 401.74$$

$$SS_1 = Std_1^2(Df_1) = (15)^2 \times 24 = 5400$$
$$SS_2 = Std_2^2(Df_2) = (25)^2 \times 19 = 11875$$

b) Calculate t:

c)

$$t = \frac{(\overline{x_1} - \overline{x_2})}{\sqrt{\frac{S_p^2}{n_1} + \frac{S_p^2}{n_2}}} = \frac{(70 - 74)}{\sqrt{\frac{401.74}{25} + \frac{401.74}{20}}} = -0.67$$

Step 6: State the conclusion:

As $t_{estimated} = -0.67 < -2.0167$ reject H_1 and there is no significant difference in test performance of Class A and Class B.

5.5 Univariate, Bivariate, and Multivariate Data Analysis

The mathematical tool used for collection, organization, and interpretation of numerical data is referred as statistics population, samples, and variable are major components of statistical analysis. Thus statistics is used to estimate the parameters from discrete or continuous variable and categorizing them as per their characteristics or patterns. Figure 5.4 shows various well-known methods for statistical analysis [7].

5.5.1 Univariate Analysis

Univariate analysis is method of analyzing data using one variable. It describes or summarizes the data and also states patterns of the data. A variable in univariate analysis states about the behavior of data or it states category of the data. Level of measurements also plays a very important role to find best fit analysis. These levels are Nominal, Ordinal, Interval, and Ratio. Data are classified according to the highest level that it fits.

A sampling of the data decides the parametric estimation of the variable and sampling can be of random, systematic, convenience, cluster, and stratified.

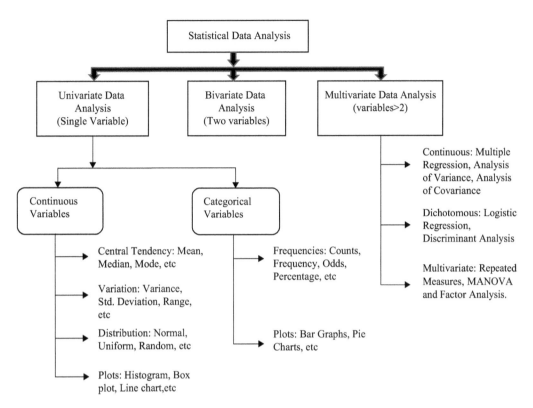

FIGURE 5.4
Methods of Statistical Analysis

Ways of expressing univariate data include central tendency (mean, mode and median) and dispersion: range, variance, maximum, minimum, quartiles (including the interquartile range), and standard deviation [8].

(i) Frequency Distribution Table

Frequency distribution table signifies the occurrence of data with its frequency in each sample of defined size. In statistics, this observation is useful in identifying the importance of data. For example, Let 0, 1, 9, 9, 2, 3, 8, 4, 6, 9, 5, 1, 1, 9, 6, 9 is sample data, then we can calculate the frequency of all symbols of data as shown in the table.

Numbers	Frequency of Numbers
1	3
2,3,4,5,8	1
6	2
9	5

These tables are useful in the categorization of data in various quantitative or qualitative vectors.

(ii) Bar Graph

It is a display tool used to present pictorial representation of data using rectangular bars as shown in Figure 5.5. Bar graphs gives quantitative information with respect to measurable parameters.

The graph plotted here shows quantity on the *y*-axis for different categories shown on the *x*-axis. Thus, comparison of various categories can be analyzed for finding low power utilization device.

(iii) Histogram

The histogram is a type of bar graph describing count of data in particular range. In a histogram, the elements in inputs are analyzed using frequency distribution by distributing these elements into "n" number of discrete bins. The histogram block sorts all complex input values into bins according to their magnitude. The histogram value for a given bin represents the frequency of occurrence of the input values. Let there are some observations of data then as per their occurrence they will be divided in a bins. For example, consider sale of 22 items is given here and quantity-wise grouping of items is required then histogram is as shown in Figure 5.6.

Items sales information = [0,1,2,2,3,4,5,6,6,7,7,8,0,8,9,10,10];

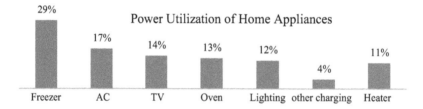

FIGURE 5.5
Bar Graph Presentation of Power Utilization

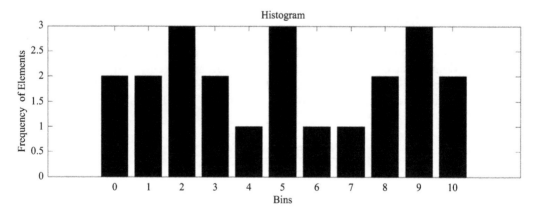

FIGURE 5.6
Histogram of Data

(iv) Pie Chart

A pie chart is a circular graph used to represent proportion or size of the category.

Each of the slices of the pie chart shows the percentage of data available in the category.

For example, Figure 5.7 shows the percentage of power utilization of home appliance and gives an idea about highly power consumable device and an economical device as per power is concerned. Pie chart is a display tool most commonly used for presenting percentage relation between various attributes.

5.5.2 Bi-Variate Analysis

The method of estimating relation or correlation or measure of association between two variables is a bivariate analysis. Normally this correlation ranges between –1 and 1, this negative or positive relationship between the variables states direction of correlation. If the correlation of variable is more away from 0 or more toward –1 and 1 then it represents more perfect the relationship between the independent and dependent variations is called degree or extent of correlation. Measures of association and statistical significance that are used may vary as per the level of measurement of the variables analyzed [5].

Flow of Bivariate Analysis:

Observe the distribution of the variable in the same group as well as in other groups of variables.

Formulate the relationship between these two distributions and find how much and in what way they differ.

In a model, if a numbers of variables are involved bivariate analysis can be applied to the standardized pairs of the variable for better performances. The methods of bivariate analysis are contingency tables, scatterplots, least squares lines, and correlation coefficients.

For example, Let there are data points describing the relation of points scored vs height of the player in basketball then a bivariate scatter plot can be used to describe their relationship as shown in Figure 5.8. It also shows regression best fit line for the estimated regression model. Persons-id is used to represent their height and score.

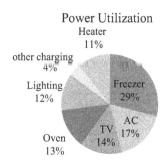

FIGURE 5.7
Pie Chart of Power Utilization

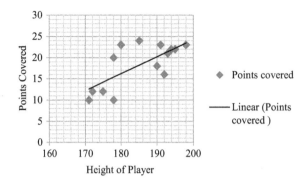

FIGURE 5.8
Scatter Plot of Points Scored versus Height in Basketball

5.5.3 Multivariate Analysis (MVA)

Most of the systems are defined with multiple independent or response variables and analysis of such system using multiple variables is a multivariate Data Analysis. This statistical technique is used to perform operable studies across multiple dimensions considering the effects of all variables on the responses of interest. The uses of MVA comprises:

a. Inverse and Capability-based design
b. Analysis of Alternatives (AoA)
c. Analysis of concepts with recent trends
d. Identification of critical design-drivers and correlations across hierarchical levels.

Multivariate analysis is suffered by high dimensionality and thus observed to be computationally costly. As most of the systems are hierarchically designed this complexity increases with its level of design. Thus in MVA parametric optimization is one of the preferred concepts of dimensionality reduction and in turn, reduces the complexity of the analysis.

5.5.4 Univariate Analysis of Variance (ANOVA)

RMS Features of EEG signal for Patient 1		
Srms	Prms	Nrms
93.13249	48.93294	41.93732
106.1351	40.69793	32.29112
103.0291	70.18156	62.07621
109.3978	82.81073	42.88057
100.5197	85.52586	25.30256
59.98254	56.242	46.18233

Univariate Analysis of Variance (ANOVA) is a statistical tool use to analyse the means of the different dataset to test variation in two or more means. It is very useful tool for analysing behavior of attributes used in classification. The technique can be used for nonspecific NULL hypothesis, feature selection, classification and prediction methodologies.

For example: Consider classification of epileptic EEG signal in Pre-seizure, Seizure and Normal state. The features are extracted such as mean, variance, Root Mean Square (RMS) for different patients. Let consider RMS feature for Pre-seizure (prms), Seizure (srms), and Normal (nrms) state then an ANOVA technique is used to identify the importance of this feature in this research by comparing prms and srms with nrms [9]. The feature table and ANOVA table is as follows:

Anova: Single Factor						
SUMMARY						
Groups	Count	Sum	Average	Variance		
Column 1	6	572.1967	95.36611	331.0251		
Column 2	6	384.391	64.06517	337.1164		
Column 3	6	250.6701	41.77835	158.8195		
ANOVA						
Source of Variation	SS	df	MS	F	P-value	F crit
Between Groups	8696.199	2	4348.1	15.77378	0.000205	3.68232
Within Groups	4134.805	15	275.6537			

Thus, if P-value is less than $\alpha = 0.05$ then the RMS features are significant.

5.5.5 Multivariate Analysis of Variance (MANOVA)

MANOVA is an extended concept of Univariate Analysis of Variance (ANOVA). In ANOVA the relation of one dependent variable is observed with respect to the group independent variables whereas in MANOVA multiple dependent variable metrics is analyzed on the basis of multiple independent variables. To address multiple dependent variables, MANOVA groups them together into a weighted linear combination or composite variable. These composite variables are known by eigenvalues, vectors, or discriminant functions.

Assumptions in MANOVA

- **Independent Random Sampling**: MANOVA assumes samples are randomly selected considering all variables are independent of one another
- **Level and Measurement of the Variables**: Independent and dependent variables are considered to be categorical and continuous variables in nature, respectively
- **The Linearity of Dependent Variables**: In MANOVA, moderately correlated dependent variables are preferred. Thus, if the dependent variables are exactly separable then the system will sacrifice on the degrees of freedom and also decreases the power of the analysis

- **Multivariate Normality**: It is very sensitive to outliers and missing value. Thus, it is assumed that multivariate normality is present in the data
- **Multivariate Homogeneity of Variance**: It also assumes that the variance between groups is equal

MANOVA statistical analysis uses multivariate F-statistics for the analysis of variable and is given by the ratio of means sum of the square and mean error for the source variable. Computation cost of MANOVA is very high as compared to ANOVA.

5.6 Principal Component Analysis

Principal Component Analysis (PCA) by Karl Pearson (1901) is a pattern recognition tool used to identify similarities and differences. PCA is a nonparametric method of extracting relevant information from complex data sets. In multivariable and high dimensions data, it is computationally difficult to analyze data thus PCA observed to be a powerful tool for analysis of multidimensional data. PCA is also a very proven method of data compression without much loss of information in data mining and also useful in predictive models, graphical representation, and regression analysis. A new set of variables are produced from PCA called principal component (PCs), which maintains a variation of original variables and are uncorrelated [10].

5.6.1 Procedure of Computing Principal Components Is as Detailed Below

Step 1: Read Data

Let **x and y are some random** vectors of population "p" then variance, covariance or correlation of "p" variables need to considered for PCA

Step 2: Subtract the mean

Calculate $(x - \bar{x})$ and $(y - \bar{y})$, where $\bar{x}$ and $\bar{y}$ are mean of x and y vectors, respectively.

Step 3: Covariance matrix calculation

Covariance matrix is calculated as shown here:

$$Cov = \begin{pmatrix} C(x,x) & C(x,y) \\ C(y,x) & C(y,y) \end{pmatrix} \tag{5.18}$$

If nondiagonal elements of the covariance matrix are positive, then both the variables are directly proportional to each other.

Step 4: Construct Eigenvectors and Eigenvalues of the covariance matrix

The eigenvectors and eigenvalues can be calculated from the covariance matrix. A scalar λ is an eigenvalue of the m × m matrix A and x is an eigenvector related to the eigenvalue λ with relation $Ax = \lambda x$. If we square the A matrix, eigenvectors remain the same and eigenvalues get squared:

$$\det(A - \lambda I) = 0 \tag{5.19}$$

Step 5: Choosing components and forming a feature vector

For the dataset of size "n", eigenvectors and eigenvalues are calculated and out of these "n" values the *highest* eigenvalue is the *principal component* of the data set. Only first few eigenvectors get selected as a final dataset with dimension "p" and feature vector can be constructed on the basis of these vectors. The feature vector is as given here:

$$\text{Feature Vector } (F) = (eig1, eig2, eig3..) \quad (5.20)$$

Step 6: Deriving the new dataset

The final dataset is derived from the formula given here:

$$\text{Final Data} = \text{Row feature vector} \times \text{row data adjust}$$

Where Row feature vector is the transpose of the feature vector and row data adjust is the mean-adjusted data transposed [10].

For Example:

Let the dataset used here contains 10 observations of 3D object movement along with three axes, then calculate new dataset representing these observations:

	X-displacement	Y-displacement	Z-displacement
	7	4	3
	4	1	8
	6	3	5
	8	6	1
X=	8	5	7
	7	2	9
	5	3	3
	9	5	8
	7	4	5
	8	2	2

(i) Derive correlation matrix

$$R = \begin{bmatrix} 1.00 & 0.67 & -0.10 \\ 0.67 & 1.00 & -0.29 \\ -0.10 & -0.29 & 1.00 \end{bmatrix}$$

(ii) As per eigenvalues concept $|R-\lambda I| = 0$ where λ can have various values as shown here

λ	Value	Proportion
1	1.769	0.590
2	0.927	0.899
3	0.304	1.000

(iii) Calculate determinant of R which is the product of the eigenvalues = λ1 × λ2 × λ3 = 0.499.

(iv) The highest eigenvalue is 1.769 considered to be a principal component and substituting it in the equation we get:

$$\begin{bmatrix} -0.769 & 0.670 & -0.100 \\ 0.670 & -0.769 & -0.290 \\ -0.100 & -0.290 & -0.769 \end{bmatrix} \begin{bmatrix} v1 \\ v2 \\ v3 \end{bmatrix} = \begin{bmatrix} 0 \\ 0 \\ 0 \end{bmatrix}$$

By solving we get feature vector:

$$F = \begin{bmatrix} 0.64 & 0.69 & -0.34 \\ 0.38 & 0.1 & 0.91 \\ -0.66 & 0.72 & 0.20 \end{bmatrix}$$

(v) Calculate transpose of *Row feature vector* ($V = F'$) and *row data adjust* ($L_{1/2}$) which is a diagonal matrix whose elements are the square root of the eigenvalues of R.

Also calculate final data by

$$\text{Final Data} = \text{Row feature vector} \times \text{row data adjust}$$

$$\begin{bmatrix} 0.64 & 0.38 & -0.66 \\ 0.69 & 0.10 & 0.72 \\ -0.34 & 0.91 & 0.20 \end{bmatrix} \begin{bmatrix} 1.33 & 0 & 0 \\ 0 & 0.96 & 0 \\ 0 & 0 & 0.55 \end{bmatrix} = \begin{bmatrix} 0.85 & 0.37 & -0.37 \\ 0.91 & 0.10 & 0.40 \\ -0.45 & 0.88 & 0.11 \end{bmatrix}$$

(vi) So 0.91 is the correlation between the second variable and the first principal component.

Next compute the communality, using the first two eigenvalues only.

$$SS' = \begin{bmatrix} 0.85 & 0.37 \\ 0.91 & 0.09 \\ -0.45 & 0.88 \end{bmatrix} \begin{bmatrix} 0.85 & 0.91 & -0.45 \\ 0.37 & 0.09 & 0.88 \end{bmatrix} = \begin{bmatrix} 0.8662 & 0.8140 & 0.0606 \\ 0.8140 & 0.8420 & -0.3321 \\ -0.0606 & -0.3321 & 0.9876 \end{bmatrix}$$

Communality consists of the diagonal elements.

$$\begin{bmatrix} \text{Var} & \text{Dia} \\ 1 & 0.8662 \\ 2 & 0.8420 \\ 3 & 0.9876 \end{bmatrix}$$

This means that the first two principal components 86.62% of the first variable, 84.20% of the second variable, and 98.76% of the third.

(vii) Calculate the coefficient matrix, B from reciprocals of the diagonals of L1/2.

$$B = VL - 1/2 = \begin{bmatrix} 0.48 & 0.40 & -1.20 \\ 0.52 & 0.10 & 1.31 \\ -0.26 & 0.95 & 0.37 \end{bmatrix}$$

(viii) Finally, the final dataset is calculated from ZB, where Z is the standard form of X and columns of matrix F are the principal factors called reduced set of variables.

$$F = ZB = \begin{bmatrix} 0.41 & -0.69 & 0.06 \\ -2.11 & 0.07 & 0.63 \\ -0.46 & -0.32 & 0.30 \\ 1.62 & -1.00 & 0.70 \\ 0.70 & 1.09 & 0.65 \\ -0.86 & 1.32 & -0.85 \\ -0.60 & -1.31 & 0.86 \\ 0.94 & 1.72 & -0.04 \\ 0.22 & 0.03 & 0.34 \\ 0.15 & -0.91 & -2.65 \end{bmatrix}$$

5.7 State Vector Machines

Vapnik and coworkers introduced supervised machine learning algorithm Support vector machines (SVMs) in the 1990s. The algorithm uses statistical learning methods of two-class discriminant functions from a set of training data. In recent research, SVM is most preferred classifier for linearly separable data and used in various applications such as bioinformatics, computer vision, handwriting recognition, text categorization, face detection and in many more prediction models.

5.7.1 Principle of Support Vector Machine

The principle of a support vector machine is to derive the optimal separating hyperplane which makes the best use of the margin of the training data. Figure 5.9 shows the construction of plot showing features and appropriate hyperplane for this dataset.

Linear SVM is a linearly scalable speedy data mining algorithm used to provide high performance solution for multiclass classification challenge. A linear SVM model has the greater performance for larger dataset with high accuracy. Support vector machine performance basically dependent on n-dimensional space of features and selection of optimizing hyper-plane[11]

Working Principle:

- Analyze the data set and perform feature extraction and feature selection then plot these features as shown in figure over n-dimensional features space
- Now iterates SVM steps for selection of optimized hyper-plane for classification of features in two classes. Let there are data points plotted with stars and circles then

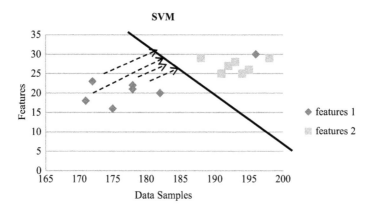

FIGURE 5.9
Principle of SVM

selection process of a hyper plane in a different situation can be observed as given below and graphically shown in Figure 5.10.

Case 1: Let there be three hyper-planes A, B, and C as shown in Figure 5.10 a then we need to find an exactly separable hyper-plane which separates two classes effectively. Here hyper-plane B is observed to be a better selection for better performance.

Case 2: A, B, and C are observed to be separating two classes of features then selection of optimizing plane will be on the basis of the distance between the plane and the nearest data points. This calculated perpendicular distance is called Margin. As shown in Figure 5.10b, it is observed that hyper-plane C has a maximum distance from data points means high margin as compared to A and B. Thus selection of plane C is better in this case.

Case 3: In this case, there are two planes having high margin but data is not exactly separable due to the nonlinear distribution of features over a space. Thus select a plane which provides high accuracy even it has low margin. Figure 5.10c hyper-plane A is observed to be a better selection B even though it has low margin.

Case 4: Here due to nonlinear distribution as shown in Figure 5.10d, exact segregation of classes is not possible then find a hyper-plane that has maximum margin ignoring uncovered outliers.

Case 5: Data distributed over a graph is fully scattered or overlapped as shown in Figure 5.10e then cannot be separated by a linear hyper-plane. Thus we need to use interrelation of x and y coordinate values of data such as $z = x^2 + y^2$ and then plot z on the graph to select hyper-plane as shown in Figure 5.10f. Here x and y values are squared and then added for z value, which will be always positive. Thus proper linear hyper-plane can be selected.

SVM is an effective function that transforms features of non-separable problem from lower space to higher dimensional space called Kernel [12]. Selection of kernel function depends on the complexity of nonlinear distribution of features and appropriate process for separation of data.

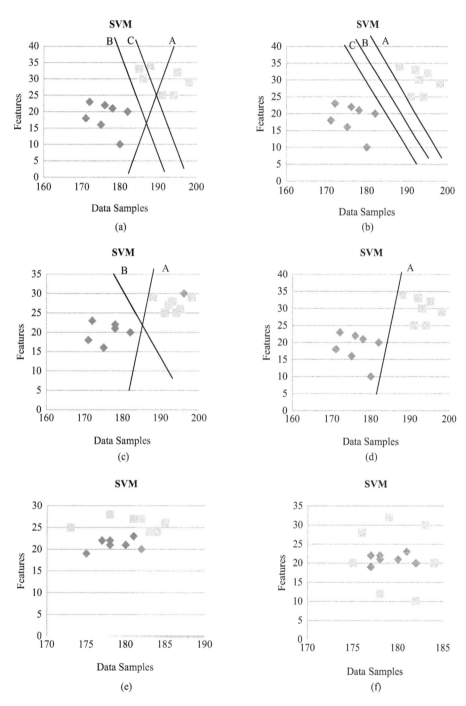

FIGURE 5.10
Selection of Hyper-Plane in Various Conditions

Advantages of SVM:

- SVM works effectively with a high margin of separation
- High dimensional spaces can be effectively addressed
- SVM is very effective if the dimension of data is larger than the sample size
- Only some of the effective training points called support vectors are used thus it is memory efficient

Limitations:

- It requires high time for classifying higher dimensional data
- Its efficiency is lower due to complex overlapping of the features
- It has a non-probabilistic interpretation for classification

5.8 Uncertainty Analysis

In statistical complex model uncertainties are observed and are named as stochastic uncertainty and subjective or epistemic uncertainty. Stochastic uncertainty is observed due to variation in system behavior at different conditions whereas epistemic uncertainty is a scientific uncertainty and it is due to limited data and related knowledge for estimation of parameters. In discrete random variables analysis, the epistemic uncertainty is modelled by probability distribution function (PDF).

For example, nuclear reactor power plant involves both stochastic and epistemic uncertainties: The risk analysis observed in a nuclear power plant states that stochastic uncertainty occurs due to the *hypothetical* accident scenarios, and uncertain parameters generate epistemic uncertainties consequently causes the estimation of the probabilities of hypothetical accident incidences.

If a system is specifically modelled for quantitative analysis then the précised quantification and propagation of uncertainty is very important[13].

In a statistics, sensitivity and uncertainty analysis are very important and performed by response surface methods as explained here:

(a) Select few highly effective parameters of the system for modelling it

(b) The sensitivity/uncertainty analysis is computed at a single point of interest

(c) The response recomputed using parameter values at the selected point of interest in (b) and the model parameters selected in (a);

(d) Construct a response surface that corresponds to the conduct of the response

(e) Estimate sensitivities and uncertainty distributions for the computed responses

5.8.1 Uncertainty Analysis Using Probability Theory

The uncertainty analysis is to evaluate the effects of parameter uncertainties on the output uncertainties. Normally the erroneousness in measured results is defined by the

Applied Statistics

uncertainty of the system if a design with confidence probability. Probability statistics is a tool to explain actual variations in the outcome of realistic observations and measurements.

5.8.2 Uncertainty Analysis of Linear and Non-Linear Systems

Mathematically the system is designed in the form of:

- linear or nonlinear relation of independent variables, calculated parameters and response variables,
- inequality or equality control that defines the values of the parameters
- single or multiple system output responses

If a system is linear system with linear dependent variables then it is represented with mathematical model "L":

$$L(a)u = Q[a(x)] x \epsilon \Omega \tag{5.21}$$

Where

1. $x = (x_1, x_2, \ldots x_n)$ is phase space position vector and Ω is a subset of real vector R
2. $u(x) = [u_1(x), u_2(x) \ldots .. u_n(x)]$ denotes linearly normalized scalar field of real vector
3. $a(x) = [a_1(x), a_2(x) \ldots .. a_n(x)]$ denotes the system parameter vector.
4. $Q[a(x)] = [Q_1(a), Q_2(a) \ldots \ldots \ldots Q_n(a)]$ denotes a vector of homogeneous sources.

The uncertainty analysis for such a system will be performed by transformations and statistics; discrete and continuous density functions; cumulative distribution functions; and indicator functions. The range of measurement error or uncertainty estimated by experimentation is a measure of the nonreproducibility. An appropriately estimated measurement uncertainty causes variation in respective result with respect to experimental results.[14]

5.8.3 Uncertainty of Measurement

Uncertainty is normally presented with range and equivalent confidence probability, and its measurement forms are classified as shown in Figure 5.11.

(i) **Natural Variability**: Natural variability observed due to temporal and spatial variability with respect to the values of the subject. Natural variability can be improved by selecting high resolution during simulation and improved calibration of model parameters. The variability of output response is fully dependent on variability of input values. The range of variation in output response varies with errors in input, the values of parameters, initial boundary conditions, model structure, processes, and solution algorithms.

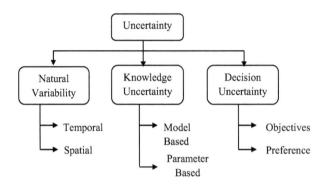

FIGURE 5.11
An Uncertainty Measurement Forms

(ii) **Knowledge Uncertainty**: It consists of parameter value and model based uncertainty affected by boundary condition uncertainty, solution algorithm uncertainty.

- **Parameter Value Uncertainty**: The estimated parameters are possible sources of uncertainty observed in output response. If the system behavior is observed by iterating model calibration procedure for multiple data sets, different parameter values would result and thus the different predictions. The prediction models are thus analyzed by parameter uncertainty with indefinite parameter values. This parameter value indistinctness is caused by fixed or variable imprecise specification of boundary conditions. These uncertainties can affect the model output, in the vicinity of the boundary, in each time step of the simulation

- **Model Based Uncertainty**: Model based uncertainty is observed due to errors in the model structure and approximations made by numerical methods used in the simulation. If the predicted model has an error then even though the estimated parameters are good, the system results in a residual model error. Normally a complex model of high dimensional space could results in errors in the model

(iii) **Decision Uncertainty**: Uncertainty in the model predictions observed by unexpected changes in model such as changes in nature, human goals, interests, activities, demands and impacts.

5.8.4 Uncertainty Analysis

Uncertainty analysis involves recognizing characteristics of probability distributions of input and output variables, and then the functions of random output variables defining performance indicators or measures [15]. An uncertainty analysis selects random input values and applies it to models to obtain the statistical measures of the distributions of the resulting outputs.

The output distributions can be used to

- represent the limit of expected outputs of the system at a certain probability level
- calculate the probability that the output will exceed a specific threshold or performance measure target value

Uncertainty analysis is often used for:

- estimating the mean and standard deviation of the outputs
- assigning a reliability level to a function of the outputs,
- describing the likelihood of different potential outputs of the system
- estimating the relative impacts of input variable uncertainties

5.8.5 Uncertainty Calculation Example

In uncertainty calculation, major parameters calculated are mean, standard deviation and probability distribution. In this example mean variation are observed for uncertainty analysis for the small dataset and large dataset with given formulas as listed in Table 5.2

Now let there be two datasets as mentioned here:

Measurements	Database 1	Database 2
X1	85	82
X2	72	81
X3	75	83
X4	80	85
X5	86	80

TABLE 5.2

Uncertainty Analysis Formulae

Parameters	Description of parameters	Small dataset	Large dataset
Mean (X_{avg})	The average value of dataset X	$X_{avg} = \frac{X1+X2+\ldots Xn}{n}$	$X_{avg} = \frac{\sum_i^n X_i}{n}$
Range (R)	It is the difference between max and min values of the dataset.	$R = X_{Max} - X_{Min}$	–
Uncertainty in A measurement (ΔX)	Uncertainty in dataset during measurement, it lies in min and max value of X	$\Delta X = \frac{R}{2} = \frac{X_{Max}-X_{Min}}{2}$	$\Delta X = \sigma = \sqrt{\frac{\sum_{i=1}^n (X_i - X_{avg})^2}{n}}$
Uncertainty in Mean (X_{avg})	Uncertainty in Mean are the values around X_{avg}	$\Delta Xavg = \frac{\Delta X}{\sqrt{n}} = \frac{R}{2\sqrt{n}}$	$\Delta Xavg = \frac{\sigma}{\sqrt{n}}$
Measured Value X_m	The final measured value in consideration of X mean and uncertainty in mean	$X_m = Xavg \pm \Delta Xavg$	$X_m = Xavg \pm \Delta Xavg$

Then we can calculate the actual value of the mean of X by considering mean of data and uncertainty in database

For Database 1:

(i) Calculate the mean by formula

$$X_{avg} = \frac{85 + 72 + 75 + 80 + 86}{5} = 79.6$$

(ii) Calculate the range of the database

$$R = X_{Max} - X_{Min} = 86 - 72 = 14$$

(iii) Calculate Uncertainty in X

$$\Delta X = \frac{R}{2} = \frac{14}{2} = 7$$

(iv) Calculate Uncertainty in mean

$$\Delta X_{avg} = \frac{R}{2\sqrt{n}} = \frac{14}{2\sqrt{5}} = 3.13$$

(v) Final Value of X mean is

$$X_m = X_{avg} - \Delta X_{avg} = 79.6 \pm 3.13 = 76.47,\ 82.73$$

Similarly, we can calculate the final value of mean for database 2 as given below

$$X_m = X_{avg} - \Delta X_{avg} = 82.2 \pm 1.11 = 81.09,\ 83.31$$

5.8.6 Software Tool for Uncertainty Analysis

UNICORN is a stand-alone software tool (http://ssor.twi.tudelft.nl/~risk/) for uncertainty analysis. The requirement of the tool is Dependence modelling of univariate random variables, although joint distributions in ASCII format are also supported.

Features of the software:

- Efficient formula parser with a min-cost-flow network solver
- Post-processing provides report generation, graphics, sensitivity analysis and probabilistic inversion

5.9 Modelling and Prediction of Performance

Modelling is a process of applying proven methods in order to analyse research problem for complexity, real world application. Modelling helped in perditions of output response depending on inputs. Research methodology normally executed in four stages as shown in Figure 5.12.

System engineering cycle is playing a very important role in solving research problem with prototype evaluation as shown in Figure 5.12. Modelling is a middle step of this cycle which exists in between system analysis cycle and system design cycle.

5.9.1 Concept of Mathematical Modelling

Real time function and methodology converted into the language of mathematics called models. As mathematics is well proven tool to design concept in consideration of assumptions, it contains well proven rules and can be easily implemented using a computer or numerical calculation.

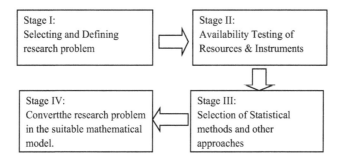

FIGURE 5.12
Execution Stages of Research Methodology

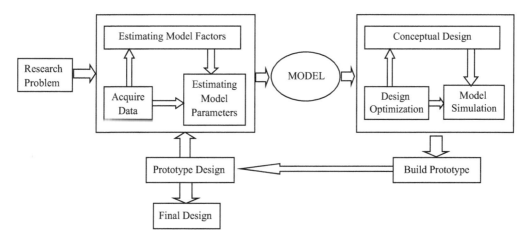

FIGURE 5.13
System Engineering Cycle

Firstly interfacing of components and approximately required variables are assumed in mathematical models then the implementation is exercised for the system. During implementation need of the parameters identified and that variable are continued further. Secondly, computation of model contributes to the complexity of the system. The basic objective of the mathematical model is to develop a scientific understanding with qualitative expression.

The overall system design flow includes understanding of research problem, selecting appropriate model for design converting it into conceptual design and verifying it with simulation, finally hardware prototyping will validate the solution to the problem. This system design cycle is explain in Figure 5.13 and elaborated as follows

5.9.2 Classification of Models

- Deterministic models produce defined outcome from the same set of input parameters
- A stochastic model determines the distribution of possible outcomes with more statistical behavior
- Mechanistic models user large data at a different hierarchical level to understand changes is the system
- Empirical models no mechanism is noted for changes but changes are analyses at different condition

5.9.3 Stages of Modelling

Modelling process is composed of four activities as shown in Figure 5.14.

(i) Building stage comprises of system analysis on the basis of assumptions & flow of the model. Choosing a suitable mathematical expression with the analogy of physical and data exploration is also covered in the building stage.

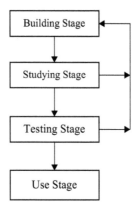

FIGURE 5.14
Stages of Modeling

(ii) Studying Model is responsible for realizing system behavior mostly qualitative behavior. Transformation into dimensionless quantities, analysing asymptotic behavior and sensitivity analysis are some of the stages of studying model.
(iii) Verification and validation of system are covered by the testing model. Testing of the system depends on.
- Assumption linearly and deterministic behavior
- Model structure
- Prediction of previously unused data
- Estimating model parameters
- Comparison of the model
- Using model will facilitate user to predict data and outcomes
- Prediction with precision estimation
- Decision support

Finally, the mathematical model will be described and tested for functional behavior of the system. Mathematical modelling is just not about quantities measures but also related to qualitative characteristics of the system.

5.9.4 Process of Modelling

The process of constructing model mostly works onto hypothesis testing [16]. The steps involved are as shown in Figure 5.15.

- Functional requirements will be studies to understand observations required for system feasible observations then forwarded to informal model

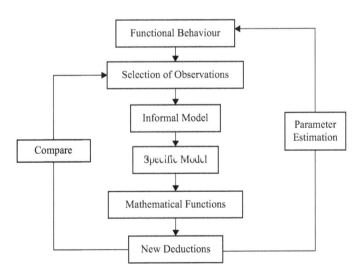

FIGURE 5.15
Process of Constructing Model

- Informal model is set at possible parameter and methods that can explain the selected observation. These methods and parameters are tested for given system to satisfy required accuracy and precision
- Formal model converts assumptions in the informal model to mathematical form. The formal model translate text based informal model in the mathematical model
- A mathematical function decides exact methodology such as logic algebra, geometry, probability theory, and computer-aided tools

5.9.5 Advantages and Disadvantage of Mathematical Modelling

Advantages:

- It decides various required characteristics of the system and defining non-feasible characteristics also
- It helped in the designing hypothesis of the system and unknown and important parameter also derived from it
- The mathematical model optimizes the system to fill up the gap between knowledge & requirement with less complexity
- The mathematical model thus can give a realistic representation of system using statistic and probable outcomes
- It is systemic, iterative and can be redefined
- It predicts behavior at a system and can help to validate methodology in a physical process

Disadvantage:

- It generates multiple outcomes of system so the selection of right outcomes is important
- As it is iterative consumes more time than traditional methods

5.9.6 Example of Mathematical Modelling: Integrated Solar Water Heater

(i) **Specification of the system**
- Data logger for analog to digital conversion with 16 bit, 0.1°C resolution
- Thermos flask
- Thermometer and thermocouples
- Software for simulation purpose
- Dynamometer to measure horizontal global solar radiations global solar radiations in w/m^2
- Leakage and pressure sensors

(ii) **Modelling solar water heater.**
Using macro model – thermal model
- Prediction of performance is a challenging task
- Define a model able to predict water temperature depending on orientation and location
- Select analytical model to predict a performance using thermal network analysis
- The macro model is given by J. I. Currie et al. [17]

Thus three parameters are considered as conduction, radiation, and convection in turns defines losses or gain.

(iii) **Macro model defines following parameter**
- Conversion from air cavity and water cavity
- Converted from cover to environment and box to environment.
- Radiation from place to cover and cover to the sky
- Conduction through the box
- The head capacitance of the system

(iv) **Computing Model To Predict Performance of System**:
Computing model of performance prediction is called performance model of the system. The performance modelling is a method of understanding performance measurements and analysis by converting characteristic of the system into mathematical form. Various international groups work onto performance model designing such as Performance Evaluation Research Center (PERC), Scientific Discovery through Advance Computation (SciDAC) Program, and Los Alamos National Laboratory [LANL].

(v) **Factors included in performance model**:
Performance modelling of a system defines with various factors & their combinations such as factors for a computer system.
- System size
- System Architecture
- Processor speed
- Multilevel cache latency and BW
- Interprocessor network latency and BW
- System software efficiency
- Type of applications
- Algorithms and paging language
- Problem size
- No of I/Os

Thus performance model can be defined to achieve accuracy, speed, latency and efficiency. The runtime functional behavior also affects the performance model.

5.9.7 Performance Measurement, Analysis, and Performance Curve

A model designed for runtime performance prediction with less computation on hardware efficient system is called performance model.

It includes steps such as: (1) Assumptions, (2) Parameters, (3) Equations
Example: Data-intensive applications designing.
Assumptions considered here are:

- Computation time is negligible as compared to I/O speed
- For work load optimization pipeline structure allows instructions to run parallel
- Input data is evenly distributed
- Deriving the model from I/O operation
- I/O Operations are mapped with equations that depend on available data

(i) Performance Measurement and Analysis

Performance measurement is an analysis of overall system including prevention and detection. Performance Measurement Process Flow can be well explained by Figure 5.16

It is just not only related to collecting data with predefined performance measures. It is the process of optimization to improve efficiency and effectiveness of the system. Most of the performance measures can be grouped into Effectiveness, Efficiency, Quality, Timeliness, Productivity, and Safety.

(ii) Performance Curve:

Various research domains concerned with classifications task uses receivers operations characteristics (ROC) curve to assess the quality of model or to compare two or more models with respect to operating points. Researchers also provide some statistics from ROC such as break-even points (BEP) or equal error rate. Instead of ROC Curve, performance curve is preferable in machine learning context.

This Expected Performance curve (EPC) provides unbiased estimates of performance at various operating points. Performance parameters normally observe for classification and some of the application specific parameters are as discussed in Example 1, Example 2 and Example 3. Example 1: Person Authentication (Model 1)

V1: False Acceptance Rate (FAR) =FP / (FN+TP)

V2: False Rejection Rate (FRR) =FN / (FN+TP)

& V: Half total error rate (HTER) = FAR/2 +FRR/2 = V1+V2/2

Example 2: Text Categorization (Model 2)

V1: Precision = TP/(TP+FP)

V2: Recall: TP/(TP + FN)

V: Aggregate Measure: 2V1.V2/(V1+V2)

Example 3: In Medical Studies

Applied Statistics

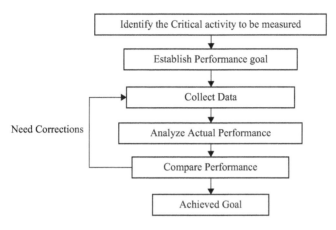

FIGURE 5.16
Performance Measurement Process Flow

V1: Sensitivity = TP/ (TP+FN)

V2: Specificity = TN/(TN+FP)

For every model two performance parameters considered breakeven point (BEP) and equal error rate (ERR) which depends on V1 & V2 for each system. Thus Expected Performance Curve (EPC) gives more realistic analysis and comparison of two models.

5.10 Multi-Scale Modelling

It is the field of solving problems which have important features at multiple scales of time or space. Multiscale modeling predicts system behavior with different models such as quantum mechanical models, molecular dynamics models, mesoscales, continuous model and device models. It allows predicting system behavior based on atomistic structure and property of elementary processes in material engineering research. In operation research, it addressed challenges for decision makers that come across organizational, temporal, and spatial scales.

5.10.1 Approaches to Analyse Multiscale Structures

- The physical analysis used for spatio-temporal structures with averaging methods, discrete method and multiscale method
- Averaging methods uses lumped parameters for certain spatio-temporal volume by assuming the system to be uniform all over the volume
- Discrete methods most commonly used in computational mechanics and chemistry as it reconstructs continuum behavior from the movement and interactions of numerous particles with discrete attributes

- The multiscale methods consider the disparity of behavior and interactions on different scales
- It identifies the prevailing behavior of complex systems
- There are different multiscale methodologies such as descriptive, correlative and analytical
  - The descriptive multiscale describes the appearance of various structures on different scales without analyzing their performance
  - The correlative multiscale methodology explains the behavior on higher scales by analysing the mechanisms on the next lower scales
  - The analytical multiscale methodology reveals the relationship between scales by formulating stability criteria dominant mechanism and compromise between submechanisms

5.10.2 Linear and Nonlinear Analysis of System

(i) **Linear System Response**:
- A mathematical model characterized by a linear relation of explanatory and response variable or load and deflection variables is called linear system
- Characteristics of the linear model
  - It can go through any disarticulation magnitudes
  - The system is very stable and has no failure or critical points in response
  - Superposition concept is applied to analyze the response of various loads
  - Linear system without load is considered to be reference system
- The requirements of such a linear model are
  - Perfect linear flexibility for any warp
  - Microscopic deformation
  - Unlimited strength
- These all requirements and characteristics are not physically realistic and also contradictory to each other. Thus a nonlinear model will be preferable. But the linear model are observed to be good in an approximation of position than the nonlinear response

(ii) **Nonlinear System**

A system in which output response is not directly proportional to the input changes means system which does not follow superposition theorem is called nonlinear system.

The behavior of the nonlinear system is described by nonlinear equations. As nonlinear equations are difficult to solve, nonlinear systems are commonly approximated by linear equations called linearization. In mathematics linear functions follows:

Additive property: f(a+b) = f(a) +f(b) & homogeneity f(2a) = 2 f(a)
When these two properties combine observed it follows superposition principle:

$$F(2a + mb) = 2f(a) + mf(b)$$

This system is observed to be linear. Nonlinear equations are normally represented with polynomial equations, for example, $a^2+b^{-1} = 0$. Nonlinear system also can be represented with differential equation such as $du/da = -u^2$ OR $du/da+u^2 = 0$.

Types of Nonlinear Response:

- Classical chaos – the unpredicted response of the system
- Multistability – Output variation in multiple states
- A periodic oscillation – a function that does not repeat values after some period [18]

Properties of nonlinear systems:

- The nonlinear system does not follow the principle of superposition
- Multiple isolated equilibrium points are observed in the system
- Every time it not possible to model nonlinear systems

Analysis methods of nonlinear system:
Nonlinear systems can be analyzed using various methods such as:

- Asymptotic techniques
- Lyapunov stability analysis
- Phase plane method
- Singular perturbation method
- Small gain theorem
- Passivity analysis

Engineering applications of Nonlinear Analysis

- Strength Analysis
- Deflection Analysis
- Stability Analysis
- Configuration Analysis
- Progressive Failure Analysis
- Envelope Analysis

5.10.3 Comparison of Linear and Nonlinear System

Sr. No.	Linear System	Nonlinear System
1	Systems output follows superposition theorem	Systems output does not follows superposition theorem
2	Outputs are directly proportional to the input	Outputs are not directly proportional to the input
3	Analysis is possible using standard test signals.	Analysis is not possible using standard test signals.
4	System do not demonstrate limit cycle	System demonstrates limit cycle
5	System do not demonstrate hysteresis/resonance	System demonstrates hysteresis/resonance

5.11 Sensitivity Analysis

The basic objective of sensitivity analysis is to measure the effects of parameter variations on output response. The local sensitivity analysis is related to analysis of output response with respect to selected parameter, points of the analysis and state variables. Whereas global sensitivity analysis will analyze critical points in the mutual phase space formed by the parameters, state variables, and adjoint variables, and subsequently analyze these critical points by local sensitivity analysis. The extent of sensitivity analysis is to analyze sensitivities of the models based on estimated parameters, around their nominal values efficiently.

Thus sensitivity analysis is involved in identifying ranges of input data for which output of linear model to remain unchanged. In statistics, sensitivity is one of tool to measure the accuracy of an instrument for classification. Thus sensitivity is defined as the probability of correctly identifying some condition.

There are two procedures adopted by statistics, viz.:

- Forward Sensitivity Analysis Procedure (FSAP)
- Adjoint Sensitivity Analysis Procedure (ASAP)

FSAP is a direct method of analysis, specifically used where the number of responses of interest exceeds the number of system parameters and/or parameter variations to be considered. ASAP is an advanced method of analysis applicable in the problem where practical interest comprises a large number of parameters and comparatively few responses.

Thus sensitivity and uncertainty analyses can be considered as formal methods for evaluating data and models because they are associated with the computation of specific parameters.

5.12 Statistica Software Tool

Statistica is an analytical tool originally developed by StatSoft for statistical analysis of data and currently provided by TIBCO Software Inc. It has a various analytical method, outputs are available in graphics, word or spreadsheet formats, tools for automation, collaboration, reporting, and data management. The software includes an array of data analysis, data management, data visualization, and data mining procedures; as well as a variety of predictive modeling, clustering, classification, and exploratory techniques. Statistica includes analytic and exploratory graphs in addition to standard two- and three-dimensional graphs.

Features of Statistica (www.statsoft.com) Tool:

- Reliable tool for descriptive and inferential statistical analysis
- Facility of selecting of multiple data mining data
- Sample data mining projects are available
- Highly flexible and customizable graphics user interface (GUI)
- Ability to handle or process multiple data streams
- Large data set is processed with optimization
- Auto response to the changes of data for analyses and results generation
- Fully programmable and customizable system
- *STATISTICA* data analysis capabilities cover thousands of *STATISTICA* functions, algorithms, tests, and methods ranging from simple break-down tables to advanced nonlinear modeling, generalized linear models, time-series methods

The tool follows design and analysis flow for execution such as data acquisition in various database formats, data preprocessing, transformation and selection, data modelling, analysis, classification or prediction, and finally reports generation in appropriate formats [19] The statistical computing in the tool is executed using "R" programming language and GNU-GPL compiler. The "R" program code can be executed and directly the outputs are available in spreadsheet or word format in STATISTICA. This tool will encourage the research for various statistical analysis for their research problem.

5.13 Summary

This chapter highlights the relation between dependent and independent variable for various inferences. It includes concept of inferential and descriptive statistics, significance of regression analysis, theory of estimation, impact of sensitivity and uncertainty, study of statistical tools and design of various linear and non-linear models. The concepts are well explained using relevant examples and applications. The chapter will contribute to researcher by stating significance of applied statistics and its usage for drawing appropriate conclusions in the research.

Further Reading

1. Vajjha Hara Gopal, "Local Information Based Parameter Estimation for Continuous Distributions", Hydrebad, India: Department of Statistics, Omania University, pp. 3–12, 2009.
2. Vijay Rohatgi and MD Ehsanes Saleh, Parametric Point Estimation. In V. K. Rohatgi & A. K. Saleh (Eds.), "An Introduction to Probability and Statistics", 2011, doi: 10.1002/9781118165676.ch8.

3. Roger Levy, "Parametric Estimation", Probabilistic Models in the Study of Language, pp. 51–76, 2012. https://www.scribd.com/document/200838920/Probabilistic-Models-in-the-Study-of-Language
4. Bud Burt Gerstman, "Basic Biostatistics", Jones & Bartlett Publishers, Berlington, 2014, ISBN-13: 978-1284036015.
5. Phil Crewson, "Applied Statistics: Desktop Reference", International and Pan-American Copyright Conventions, US, 2016.
6. Vogt W. Paul, Burke Johnson, "The Dictionary of Statistics & Methodology: A Nontechnical Guide for the Social Science", SAGE Publications, USA, 2011.
7. http://gchang.people.ysu.edu/class/s5817/L/Diagram02.pdf
8. Stephanie, "Univariate Analysis: Definition, Examples", Statistics how to, Word Press, 2018.
9. Varsha Harpale, Vinayak Bairagi, "An Adaptive Method for Feature Selection and Extraction for Classification of Epileptic EEG Signal in Significant States", Journal of King Saud University - Computer and Information Sciences, Elsevier, King Saud University, 2018, ISSN 1319-1578.
10. Jack Prins, "Process or Product Monitoring and Control". In NIST/SEMATECH e-Handbook of Statistical Methods, 2012, http://www.itl.nist.gov/div898/handbook/pmc/section5/pmc552.htm.
11. Alessia Mammone, Marco Turchi, Nello Cristianini, "Support Vector Machines", WIREs Comput. Stat., vol. 1, no. 3, (November 2009), pp. 283–289, 2009, doi: 10.1002/wics.49.
12. https://www.analyticsvidhya.com/blog/2017/09/understaing-support-vector-machine-example-code/cs229.stanford.edu/notes/cs229-notes3.pdf
13. Jorge Arturo Hidalgo Toledo, "Model Sensitivity and Uncertainty Analysis". In Water Resources Systems Planning and Management, pp. 255–290, 2009, ISBN 92-3-103998-9. https://issuu.com/jhidalgo54/docs/09_chapter09
14. Lab Manual, "Averaging, Errors and Uncertainty", Department of Physics and Autonomy, University of Pennsylvania.
15. Dan Gabriel Cacuci, "Sensitivity and Uncertainty Analysis, Data Assimilation, and Predictive Best -Estimate Model Calibration", Handbook of Nuclear Engineering, Springer, Bostan, MA, pp. 1913-2051, 2010. https://doi.org/10.1007/978-0-387-98149-9_17.
16. Philip Schrodt, Mathematical modeling. In J. B. Manheim, R. C. Rich, & L. Willnat (Eds.), "Empirical Political Analysis: Research Methods in Political Science", New York: Longman, pp. 290–312, 2002.
17. J.I. Currie, C. Garnier, T. Muneer, T. Grassie, D. Henderson, "Modelling Bulk Water Temperature in Integrated Collector Storage Systems", Building Services Engineering Research and Technology, vol. 29, no. 3, pp. 203–218, 2008.
18. Klaus-Jürgen Bathe, Introduction to Non-Linear Analysis. In "RES.2-002 Finite Element Procedures for Solids and Structures", Spring, Massachusetts Institute of Technology: MIT OpenCourseWare, https://ocw.mit.edu. License: Creative Commons BY-NC-SA, 2010.
19. S. Sarumathi, N. Shanthi, S. Vidhya P. Ranjetha, "STATISTICA Software: A State of the Art Review", World Academy of Science, Engineering and Technology International Journal of Computer and Information Engineering, vol. 9, no. 2, pp. 473–480, 2015.

6
Presenting and Publishing the Research Findings

Krishna Warhade, Vinayak K. Bairagi, and J. Jayanth

CONTENTS
6.1 Introduction 208
6.2 Types of Publication 209
6.3 Body of Research Papers, Reports, and Theses 210
    6.3.1 Body of the Research Paper 210
    6.3.2 Body of Reports and Theses 211
6.4 Contents of Research Paper, Reports, and Theses 212
    6.4.1 Title 212
    6.4.2 Author's Name and Affiliation 212
    6.4.3 Abstract 212
    6.4.4 Introduction of Paper Contents 213
    6.4.5 Literature Review 213
    6.4.6 Methodology 214
    6.4.7 Experiment Results and Discussions 214
    6.4.8 Conclusion and Future Scope 215
    6.4.9 Acknowledgment and Conflict of Interest 215
    6.4.10 References 216
    6.4.11 Tables and Figures 217
    6.4.12 Authorship 217
6.5 Review System 218
    6.5.1 Types of Peer Review 218
6.6 Guidelines for Selecting the Journal for Publishing Paper 219
6.7 Identifiers to Identify Journals and Research Papers 220
6.8 Subscription-Based Journal and Open Access Journals 220
6.9 Journal Ranking and Journal Metrics 221
    6.9.1 Journal IF (Impact Factor) 221
    6.9.2 Source Normalized Impact per Paper (SNIP) 221
    6.9.3 SCImago Journal Rank (SJR) 222
    6.9.4 Cite Score 222
    6.9.5 Eigen Factor and Article Influence Score 222
    6.9.6 H Index and I10 Index and G Index 222
    6.9.7 Research Gate (RG) Score 225
6.10 Citation Index, Databases, Search Engines, Author Identifiers 225
    6.10.1 Citations and Citation Index 225
    6.10.2 Most Preferred Citation Index and Search Engines 226
    6.10.3 Some Other Indexing Services 226
    6.10.4 Unique Author Identifier for Efficient Search 226

        6.10.5 Online Free Repository 228
6.11 Use of Documentation Tools, Bibliography Tools, and Presentation Tools
        Useful for Writing and Presenting Paper and Theses. 228
6.12 Presenting Research Papers, Reports and Theses 229
        6.12.1 Preparation and Presentation. 231
        6.12.2 Presenting Posters 232
6.13 Summary 234
Further Reading. 234

Put it before them briefly so they will read it, clearly so they will appreciate it, picturesquely so they will remember it and, above all, accurately so they will be guided by its light.

—Joseph Pulitzer

Learning objectives of this chapter include:

- Provide guidelines about contents and body of research papers, reports, and theses
- Signify importance of various journal identifiers and author identifiers
- Introduce various journal metrics with its significance
- Create awareness about various citation index, databases, and search engines
- Introduce standardized approach of documentation, bibliography, and presentation tools
- Guide way of effective present of research papers, posters, and theses

This chapter will enable the researchers to

- Write research papers, reports, and theses effectively in a proper flow with quality contents
- Understand definition of various metrics and the selection of referred journals
- Effectively use various search engines and citation index for research
- Recognize authenticity and originality of published journal using ISSN, DOI and use ORCID, SCOPUS, and Researcher ID effectively
- Choose documentation and bibliography tools for writing quality papers, reports, and theses
- Present research papers, posters, reports, and theses in a stipulated time period

6.1 Introduction

Presenting the research and publishing it, is one of the essential ingredient of research. After rigorous and scrupulous experimentation, it is always rewarding to share the research findings with the research community and it may be looked as a moral responsibility as well. Publishing a paper in a reputed journal / conference is a challenging demanding apt and precise presentation of task obtained results.

In today's scenario, researchers prefer disseminating their novel contributions on various reputed platforms like book chapters, newsletters, conferences, and journals and are keen to increasing their citations. A reputed publication necessarily ensures thorough knowledge of researchers, contributions and innovations leading to advancements in the scientific community. Such a recognized publication automatically benefits not only the author/ researcher but also serves as a torch bearer to the entire research community and scientists with similar interests.

Good publication is also sometimes looked as a key to career advancements, gaining monetary benefits. Writing a clear, and a reader friendly manuscript that highlights novel findings /contributions, supports it with valid results and signifies its importance, is critical and inevitable component of a good publication.

Writing /drafting of scientific manuscript is less about the individual style of the author and more about the standard style recommended by the respective journal / conference.

The section *'instructions to the authors'* on the journal/conference website details the expected format of the manuscript.

Nevertheless, an impressive manuscript , should necessary include :

- Detailed research methodology and specific comparison of the obtained novel results/contributions with the reported literature
- Reviews of the work related to particular area of interest
- Documents that advances the knowledge of the recent trends in the area of interest

Following points should be avoided while publishing papers in scientific publication:

- Documents not related to scientific interest
- Documents which are not in trend (outdated)
- Replica/duplication of the already reported research
- Unacceptable methodology, results, and conclusion

You need a GOOD manuscript to present your contributions to the scientific community.

6.2 Types of Publication

Full articles/original articles/research articles are often substantial and significant completed pieces of research.

- The page length of these articles is typically four to six printed pages
- It requires sufficiently high quality of the paper, representing the novel methodology and execution of the work
- It presents novel results, fulfilling the expectations of the reviewers and editors with expertise and awareness of recent trends in that domain
- It requires significant contributions in comparison with reported literature

Letters/rapid communications/short communications are quick and early communication of significant and original advances. They are much shorter than full articles:

- They have short page length
- The review time is less as compared to a full-length article
- They usually require original results with significant change in the technology
- Routine small incremental results are not suitable for this letter category of publication

Review papers/ perspectives details the ongoing developments since the origin of the problem and summarizes reported results. It is not a platform to introduce new information:

- Review papers provide a comparison of two or more theories and evidence of each work
- Review papers should also provide development of new tools, methods, and theories
- Work should also compare various methodologies for same problem with their pros and cons
- A novel insight gained from a wider view of recent progress on a topic, or the recognition of a critical new problem or issue previously unnoticed
- A controversy: two or more camps with competing theories or explanations of a phenomenon, with evidence for each

Various types of research publications are shown in Figure 6.1.

6.3 Body of Research Papers, Reports, and Theses

Research paper starts with introduction of the domain, defined problem statement, followed by detailed literature survey. The main section of the paper is the methodology, proposed algorithm, experimental results, discussion and conclusion.

6.3.1 Body of the Research Paper

Broadly, the body of the research paper usually consist of,

- Title
- Authors name and affiliation
- Abstract and keywords
- Table of contents
- Introduction
- Methodology and proposed algorithm
- Results and discussions
- Conclusion

Presenting and Publishing Research Findings 211

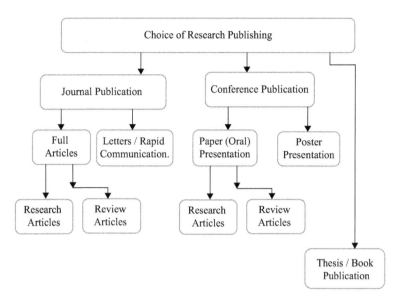

FIGURE 6.1
Various Types of Research Publication

- References
- Appendices, if any

6.3.2 Body of Reports and Theses

Reports and theses are broadly categorized in to three parts

(i) **Preliminary pages includes:**

Title page, declaration by researcher, certificate by supervisor, abstract, table of contents, list of figures, list of tables, symbols, acronyms, and their meanings.

(ii) **Main body of the theses includes:**

The body includes following elements:
- Introduction
- Literature survey
- Methods used and proposed algorithms
- Results
- Discussion
- Conclusions

(iii) **Acknowledgments and reference material**

It includes these elements:
- Acknowledgments
- Bibliography
- Appendixes

6.4 Contents of Research Paper, Reports, and Theses

An imposing manuscript should be presented clearly, must be easily understandable, and should convey the logic and scientific message to the researchers. The structure of the full article is title, authors, abstract, keywords, main text with introduction, methods, results and discussion, conclusions, acknowledgments, references, and supplementary material.

6.4.1 Title

Title must be short and meaningful. It should give an idea of the research problem addressed/ presented in the manuscript. Research papers are categorized/searched by the keywords in the title. It is therefore imperative to be careful about the choice of the keywords in the title.

The title should be informative and concise without any abbreviations. The title of the manuscript leaves the first impression to the reviewers/editors and should necessarily justify the subject matter or content of the manuscript. Mismatch in the same is likely to lead to direct rejection of the manuscript Keywords are the specific labels used in the manuscript and should use only establish abbreviations related to the title.

For Example:

- Poor title: A study of experimental tube with solar collector using supercritical CO_2
- Correct title: An experimental study on evacuated tube solar collector using supercritical Carbon dioxide
- Keywords: Solar collector; supercritical CO_2; solar energy; solar thermal utilization

6.4.2 Author's Name and Affiliation

A team of researchers claims/owns the authorship of the manuscript, based on their individual contributions. Such a team may include corresponding author, mentor/ supervisor/guide/co-supervisor/co-guide and other team-mates who have affirmatively contributed in the research and the writing of the manuscript. The affiliations of the authors and their memberships of the professional bodies, if any, are necessary in the manuscript.

6.4.3 Abstract

Abstract of the manuscript is one of its most important component. It summarizes the entire research detailed in the manuscript using mere 200-250 words. The choice of words in the abstract should therefore, be very impressive and should effectively present a bird's view of the manuscript. Abstract of the paper points to the research domain and the importance of the addressed problem. It states the gaps in the reported literature and further emphasizes on the addressed issues. A striking abstract indicates the methodology/ novel approach used to obtain the claimed results and precisely, presents its efficacy, highlighting the contributions of the research. On a lighter note, the abstract of the manuscript is equally important and impactful as the trailer of the movie. It indeed, advertises the manuscript.

- Abstract should be limited to 200–250 words
- One to two sentences should specify the aim and the problem addressed in the manuscript
- Two to three sentences should contain materials and methods used
- Two to three sentences should emphasize the novelty of the results which should be compared to the reported literature
- Abstract should conclude with an argument signifying the contribution

Check the guidelines of the journal or conference for specific expectations regarding the abstract. Abstract should be always be extremely impressive, should summarize the entire manuscript and finally and leave a punch on the reader. Refer to [1] for details.

6.4.4 Introduction of Paper Contents

Main objective of introduction is to describe the purpose, motivation, relevance and importance of research, to the research community. It is a key paragraph for readers. The reader will be more interested in a paper, if the introduction is clear and well organized. Researchers who are unfamiliar with subject or domain of presented paper should be able to read article quickly and understand the condensed history of the domain. Hence it is important to write introduction and history of the problem statement with citations. Mention the necessity of research and its significance in this part. Start with few sentences that introduce topic to reader and make these sentences more interesting. In the paper, the introduction should be limited to few pages whereas in theses you can write one entire chapter on introduction. The sequence in the introduction part must be logical. Present the contribution and studies reported by the other researchers in the domain with latest development in the topic. Outline the research statements before starting the initial draft. Present the initial draft with logical linking of paragraphs. Once the paper or theses draft is ready, revise your introduction before finalizing the further content. Some researchers opt to write introduction part after writing the full draft of theses. Never include any results and findings in the introduction.

6.4.5 Literature Review

Literature survey should be well structured and logically connected. It presents the overview and current status of the problem statement. Include the literature which has significantly contributed to the domain. Link the relationship of each finding to one another. Relate new finding to the previous findings. Select the most relevant literature for your research problem. It should focus on the relevant theories put together by the experts in the field. Search for reliable and recent material on the topic. You should try to review major books and reputed journals for the quality content. Try to include all existing relevant research in your topic subject to limitation of pages in paper and theses. Number of books and journal articles to be reviewed depends on level of study and type of research. During review, give more importance to the reputed journals papers in your field of work. If you are writing a Master's theses, the literature review can be around 2,000–3,000 words, whereas for a PhD theses it may be one entire chapter. Traditionally, the literature review is organized in historical order. In conceptual format, literature is organized according to various theories. Literature review can also be organized methodologically. Be selective and precise in summarizing the results and drawing conclusions from the available literature and use own words to summarize it. Identify the gaps and describe it clearly. Compare and evaluate research issues, point to unidentified problems in the domain. Literature survey should focus on finding and summarizing ongoing research since the origin of the problem. The ultimate goal of the literature survey should be to critically analyze research, identify gaps, and propose solutions; refer to [2] for details.

6.4.6 Methodology

Methodology should include information about laboratory setup, equipment used; primary or secondary data used that will allow to validate and further improve results. Methodology must also include statistical models or own methodologies used to analyze the data. Some research problem demand use of publically available datasets and sometimes practical datasets may also have to be created using reliable and acceptable sources. Ensure that you are using large number of samples to be able to analyze effectively, prove the efficacy and finally recommend the proposed approach. Unreliable methods produce unreliable results and may lead to wrong interpretation.

The methodology section may be presented in the following sequence:

- Discuss overall approach for investigating the research problem
- Proposed methodology must be suitable to achieve objective of the research
- Describe the methods used for data collection. If practical data is created, describe the sources of the data and justify its reliability for acceptance by the research community
- Explain the statistical method is used for data analysis. If new algorithm is proposed then explain the theoretical concept and your contribution clearly
- Explain limitation, if any, of data collection method used by the researcher. Process of collecting data greatly impacts the findings. This will help other researchers to take car while extending your research. Avoid using irrelevant and unacceptable data
- Explain the approach used to study the defined/identified problem
- If a novel approach is proposed/presented, feature the micro-details of the approach to enable revalidation by other researchers
- If an already reported method is adapted and further modified/extended, cite the original idea and showcase the results achieved due to the proposed modification. Provide appropriate reasoning and justification for the obtained improvement and contributions made
- Specify the hardware/software components and the system configurations used for implementation
- Use of acceptable and uniform notations to ensure easy understanding of your study

6.4.7 Experiment Results and Discussions

In experimental results, describe following points initially:

- Which data is used and how much portion of the data is used for analysis. If secondary standard data is used, provide details

- Which methodology, statistical method or algorithm is used for analysis of data from methodology section?
- Which instrument and/or software is used for implementation of algorithm or methods?
- Provide specification and configuration of instrument and PC/laptop used

Do not entirely rely on descriptive text, instead use figures, charts, and tables to present results effectively. Select tables and figures that show research findings and significant contributions clearly. Highlight the important findings discovered. Focus only on findings that are important and related to the research problem. State the limitations of the research findings clearly. Highlight negative results emerged from the study.

In the 'discussion' section of the manuscript/ theses, researcher should:

- Include major findings of the research
- Explain importance of findings and how they are crucial to your research area
- Provide relevance of the obtained results to similar studies
- Discuss the research contribution and limitations
- Suggest extension pointers for further research
- Avoid exagaration / mis-interpretation of achieved results or critization of reported ones

6.4.8 Conclusion and Future Scope

Conclusion should be the best part of the theses and paper. In addition to overall conclusion, each chapter of the PhD theses should also have a conclusion. In journal paper, conclusion should be restricted to one paragraph.

The conclusion should discuss findings of the research. In addition to significant contribution, highlight the limitation of the algorithm or research findings. This will help other researchers to improve the limitation of your algorithm In the 'conclusion' section of the manuscript/ theses, researcher should:

- Not repeat the sentences in the abstract and results
- Discuss the improvement / contribution /significance of your study as compared to the reported literature or recent developments
- Suggest future scope of the study and provide pointers to further improvement

6.4.9 Acknowledgment and Conflict of Interest

- Acknowledgment section should express thankfulness, help received and gratitude towards everyone who have contributed in the completion of the study and success of the research
- The statements should be professional, brief, and more informal than rest of the text, but specific to the people who have helped in completing the work. Example: I would like to thank Professor. Dr. Shivaprakash Koliwad, MCE Hassan, Karnataka for his expertise and advice in the completion of this project

- This section includes a thanking note for all the supporters and members who have offered any direct/indirect assistance , technical/non-technical help during the research. Example: I would like to thank Dr. Ridhima Pandit, Dept. of English, HC College, Maharashtra for her extraordinary efforts in drafting the manuscript and ensuring appropriate use of English language to reach a larger community of readers
- In cases of funded research projects, financial help received from various local/ national/ international funding agencies or scholarships / fellowships received should be acknowledged. Example: This work is supported by Department of Science and Technology, DST, Government of India under research grant: SR/TP/ETA-15/2009
- List other helpers who had helped in the laboratory experimentation, other colleagues and family members. Example: I would like to thank my friends and colleagues for their contribution in the labs. I would like to thank my family members for their support

Conflict of interest is a situation in which the authors are interested in multiple concepts and serving one interest would be; like working against another. It is necessary to declare the statement conflict for few referred journals.

6.4.10 References

References are detail description of the papers, books, or other sources from which researcher has obtained the information. Researcher must acknowledge the sources used, which will enable other researchers to identify and trace the sources for their research. References contain only those work cited within text. The bibliography includes not only the text referred in paper, but also the additional background material required for the reader.

The reference style is generally mentioned by universities for theses and by journals for papers.

The major citation styles and their examples include:

- IEEE

K. Warhade, S. Merchant and U. Desai, "Effective algorithm for detecting various wipe patterns and discriminating wipe from object and camera motion", IET Image Processing, vol. 4, no. 6, p. 429, 2010.

- APA (American Psychological Association)

Warhade, K., Merchant, S., & Desai, U. (2010). Effective algorithm for detecting various wipe patterns and discriminating wipe from object and camera motion. *IET Image Processing, 4*(6), 429. doi: 10.1049/iet-ipr.2009.0301

- Chicago Manual of Style

Warhade, K.K., S.N. Merchant, and U.B. Desai. 2010. "Effective Algorithm For Detecting Various Wipe Patterns And Discriminating Wipe From Object And Camera Motion". *IET Image Processing* 4 (6): 429. doi:10.1049/iet-ipr.2009.0301.

- MLA (Modern Language Association)

Warhade, K.K. et al. "Effective Algorithm For Detecting Various Wipe Patterns And Discriminating Wipe From Object And Camera Motion". *IET Image Processing*, vol 4, no. 6, 2010, p. 429. *Institution of Engineering And Technology (IET)*, doi:10.1049/iet-ipr.2009.0301. Accessed 4 Aug 2018

- Turabian Style

Warhade, K.K., S.N. Merchant, and U.B. Desai. "Effective Algorithm For Detecting Various Wipe Patterns And Discriminating Wipe From Object And Camera Motion". *IET Image Processing* 4, no. 6 (2010): 429. doi:10.1049/iet-ipr.2009.0301.

- Harward

Warhade, K., Merchant, S. and Desai, U. (2010). Effective algorithm for detecting various wipe patterns and discriminating wipe from object and camera motion. *IET Image Processing*, 4(6), p.429.

6.4.11 Tables and Figures

Tables present data directly and are preferred over graphs when the exact numerical values of the data are needed. It is easier for a reader to compare numbers arranged within a row than within a column. It is also easier to compare numbers that are close to each other. Often, 2D tables will benefit from marginal analysis, where rows and columns are totalled or expressed as a percentage of the total. Different journals have different formatting requirements for tables. For example, many journals allow only horizontal lines in the table. Before submitting the manuscript, review the table formatting guidelines and render the table in the required format.

As a form of communication, figures (and in particular, the graphical display of quantitative data) are uniquely suited to convey information from complex data sets; quickly and effectively. Whereas statistical analysis aims for data reduction (expressing a mass of data by a few simple metrics), graphing retains the full information of the data.

Guidelines for figures:

- Figure captions should not be an afterthought, they are an integral part of the figure
- Label within the graph or in the caption as necessary to minimize the need to refer back and forth from the text
- Graphs should have a title. Put the title information in the figure caption
- Pie charts are not always the best option. Explore other options as well
- Use bar charts only when you cannot find a better option. Bar charts should only be used to plot categorical data, but if the categories have a natural order, then a line plot will usually work better
- Avoid all spurious three-dimensional (3D) effects, such as the use of 3D bars in a bar chart. They only lead to confusion, never to greater clarity
- Choose plot scales (x- and y-axis start and stop values, for example) to avoid white space: try to use at least 80% of each scale to display data

6.4.12 Authorship

An author of a scientific paper is anyone who has made a creative contribution to the words or ideas being presented that are claimed to be novel. To determine the proper list of authors for a paper, first ask, "What is novel about this work?" Then ask, "Who contributed to the creation of this novel content?"

Authorship gives both credit and responsibility for the work. In this case first author is assumed to be the person whom most credit and responsibility accrue. Authors are then ordered according to decreasing contribution to the work. But different communities have different cultures, and this system of author ordering is not universal. Another system is quite common when publishing involves the work of PhD students or postdocs. Here, the work generally represents the theses project of one student, who is then assigned the first author position. That student's supervisor is assigned the last author position. There are two undesirable approaches of listing the authors for a manuscript: leaving off someone who belongs on the list (a *ghost* author) and including someone who does not belong on the list (a *guest* author).

6.5 Review System

The sequential processes and steps involved in the review systems are detailed as follows:

- The manuscripts is initially screened and checked to match the guidelines provided by the journal. The manuscript is examined for plagiarism. Authors will be informed within one to two weeks after successful plagiarism check
- The senior editor decides whether to proceed or if manuscript should be rejected without review. It is informed within 10 days after plagiarism approval
- Editors identify the reviewers depending on the expertise and their reference databases which are constantly being updated
- Typically the manuscript will be reviewed within 90 days. Should the reviewers' reports contradict one another or a report is unduly delayed, a further expert opinion will be sought
- The Senior Editor is responsible for the decision to reject or recommend the manuscript for publication. This decision will be sent to the author along with any recommendations made by the referees
- If changes are required 30 to 60 days' time will be provided for resubmission and again it will be verified by the editor and reviewer
- After acceptance, paper will be typeset and will be sent back to author for proof reading. After proof reading and final corrections, the article is made available online (online first edition), and a few weeks later this is compiled into an online volume and issue

6.5.1 Types of Peer Review

Peer review is the process of evaluation of scientific work by others who are expert in the same field. There are various types of peer review process as follows

Presenting and Publishing Research Findings 219

- **Single-blind review**

In this type of review, the names of the reviewers are hidden from the author. Reviewer knows about author of paper but author is unaware about the reviewer of his/her manuscript

- **Double-blind review**

In this model of peer review system, reviewers don't know the author of the paper and the author does not know the reviewer of his/her paper

- **Triple-blind review**

With triple-blind review, reviewers are anonymous and the author's identity is unknown to both the reviewers and the editor

- **Open review**

In this model of peer review, both the reviewers and authors are known to each other during the peer review process

For Example: Reviewers scheme of Taylors and series is as shown in Figure 6.2.

6.6 Guidelines for Selecting the Journal for Publishing Paper

To select the suitable journal first understand research area or domain and its target application. It is suggested to follow some consideration before selecting any journal

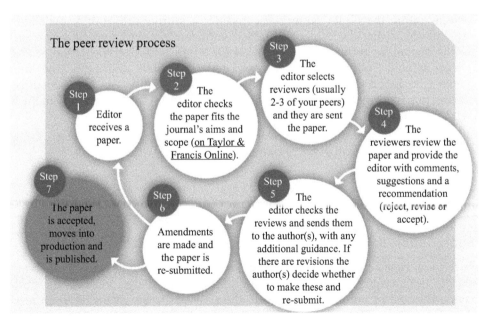

FIGURE 6.2
Peer Review Process followed by Taylor and Francis Publishers (Courtesy: Taylor and Francis Publishers)

for publication of the research paper or article. Many universities have set up the guidelines for publishing papers and have provided the list of journals where research scholar is expected to publish the paper. General expectations are, the journal should be listed in Science Citation Index (SCI) or Scopus Index list and having impact factor from Thomson Reuters (Now Clarivate Analytics), however there are few high quality journals to whom impact factor is not provided though they are quite old (i.e., *Journal of the Institution of Engineers*, established in 1920 and granted royal charter in 1935). Hence, take the opinion of the research supervisor before submitting the manuscript to any journal. One must visit the journal home page for checking the scope of journal. However, there are some website for the guidance to select the suitable journal [3–9].

6.7 Identifiers to Identify Journals and Research Papers

There are various identifiers/numbers/code formats available to recognize authenticity and originality of published journal. Example- ISSN/DOI/PMID/SSN number:

- **ISSN:** The term International Standard Serial Number could be a distinctive variety to determine a print or electronic periodical (journal) title. ISSN numbers are usually encoded in BARCODE formats and are available on journal itself and/or on publisher's websites, or on 'ISSN.org' for all cases. Moreover, there are some other online websites available for validating or verifying ISSN numbers authenticity and provide necessary information to the author [10–13]
- **DOI:** Digital Object Identifier (DOI) is employed to unambiguously determine objects within the digital setting, for instance a journal article or knowledge set. An Interior is that the symbol of varied entities viz physical, digital or abstract. DOI's syntax is outlined by ISO 26324:2012 info and documentation – Digital Object symbol System (DOI). It's composed of a prefix and suffix: the interior prefix is assigned to a company by an Interior Registration Agency; the suffix for an Interior is made by the organization depositing the Interior for a content item within the Interior system
- **PMID:** PubMed Identifier (PMID) is a unique number allotted to each PubMed citation. PubMed is collection of biomedical literature

6.8 Subscription-Based Journal and Open Access Journals

In subscription based journal the readers are paying for article to read it, whereas the authors are not charged for publication charges. In open access journals the authors are paying the cost of publication, whereas the article is freely available to all readers without any charges. There are more chances of getting referred by other researcher as the article is freely available to read and to download. Open access journals are funded by article processing charges paid by the research sponsors. Some of open access papers are funded by academic institution, learned society where publications charges are not paid by the authors. Some of open access papers are provided access after 6–18 months.

6.9 Journal Ranking and Journal Metrics

Journal ranking is widely used in academic circles in the evaluation of an academic journal's impact and quality. Journal rankings are intended to reflect the position of a journal in a specific field/domain, the relative difficulty of being published in that journal, and the prestige associated with it. Journal ranking is based on indexes/abstracts given by ISI, SCOPUS, DBLP, Google, and so on.

There are a various types of journal metrics. Journal metrics are used in the evaluation of impact and quality of journal. The most usable metrics are:

- Journal impact factor
- SNIP and SJR and CiteScore metrics
- EigenFactor and Article Influence Score
- i10 index
- H index
- Immediacy Index along with cited half-life
- RG Score

6.9.1 Journal IF (Impact Factor)

The journal Impact Factor (IF) is revealed each year by Thomson Reuters (now owned by Onex Corporation and Baring Private Equity Asia, and known as Clarivate Analytics). The factor may be a quantitative relation of total range of articles cited in the last two years and total range of articles published in the last two years. If 2017 is the current year, then journal impact factor of 2017 is calculated by the formula:

Journal Impact factor = [Citation 2016 + Citation 2015] / [Publications 2016 + Publications 2015]

Impact factor definition is evolved over the years [14]. The impact factor of reputed journals from IEEE, Springer, Elsevier, and Taylor and Francis are measured by Thomson Reuters.

One can find Journal citation report released by Clarivate Analytics which includes journal covered by Science Citation Index (SCI), Science Citation Index Expanded (SCIE) and Social Science Citation Index SSCI journals [3,15].

Few websites are useful to find impact factor of a particular journal [16], one can search by the name of journal, title, or ISSN.

6.9.2 Source Normalized Impact per Paper (SNIP)

Some categories of subject are receiving more citation as compared to others. So measuring only Impact Factors of journal may not be sufficient in some cases of comparing the two journals of different domains. SNIP (Source Normalized Impact per Paper) is the kind of normalization process to normalize the difference in citation potential of different subjects. The process of normalization includes decreasing the raw impact factors of journals in subjects with a higher frequency of citations whereas increasing the raw impact scores for journals covering subjects with relatively fewer citations. It is ratio of impact factor of particular journal to the relative citation potential of that journal in specific domain.

6.9.3 SCImago Journal Rank (SJR)

Is primarily based on citation facts of the more than 20,000 peer-reviewed journals listed through Scopus from 1996 onward. Citations are weighted, relying at the rank of the mentioning journal. It expresses the average number of weighted citations received in the selected year by the documents published in the selected journal in the three previous years [7]. A weight is given to the citation based on the status of the journals where such citations come from.

6.9.4 Cite Score

CiteScore is a new metric introduced by Elsevier which gives a more comprehensive, transparent and current view of a journal's impact. It is the ratio of the citation received in current year to the paper published in last three year to the number of documents published in last three years. Few websites will help in finding the CiteScore, SNIP, and SJR of a particular journal [17]. For example if the site is searched by journal title IEEE Transaction on image processing, CiteScore, SNIP, and SJR will be provided (refer to Figure 6.3 for the screenshot).

6.9.5 Eigen Factor and Article Influence Score

The Eigen factor score is, developed by Jevin West and Carl Bergstrom at the University of Washington. Journals are rated according to the number of incoming citations, with citations from highly ranked journals weighted to make a larger contribution to the eigen factor than those from poorly ranked journals. Eigenfactor scores and Article Influence Scores are calculated by eigenfactor.org

The Article Influence Score determines the typical influence of a journal's articles over the primary five years after publication.

AIS = (0.01* Eigen factor score) / X

Where X is 5 year article count divided by the 5-year article count from all journals

6.9.6 H Index and I10 Index and G Index

- **H-index:** The h-index is a metric for evaluating scientist's personal performance based on publication and citation. It calculates scientist's performance primarily based on their profession publications, as measured through the lifetime wide variety of citations each article gets. The dimension is dependent on quantity of publications and range of citations of an individual researcher. H-Index is utilized as most mainstream reference check record, it was concocted by Jorge Hirsch in 2005, here h-Index's "H" implies Hirsch, and it is a numerical pointer of how gainful and powerful a specialist is. The h-record of a researcher is the number "n"

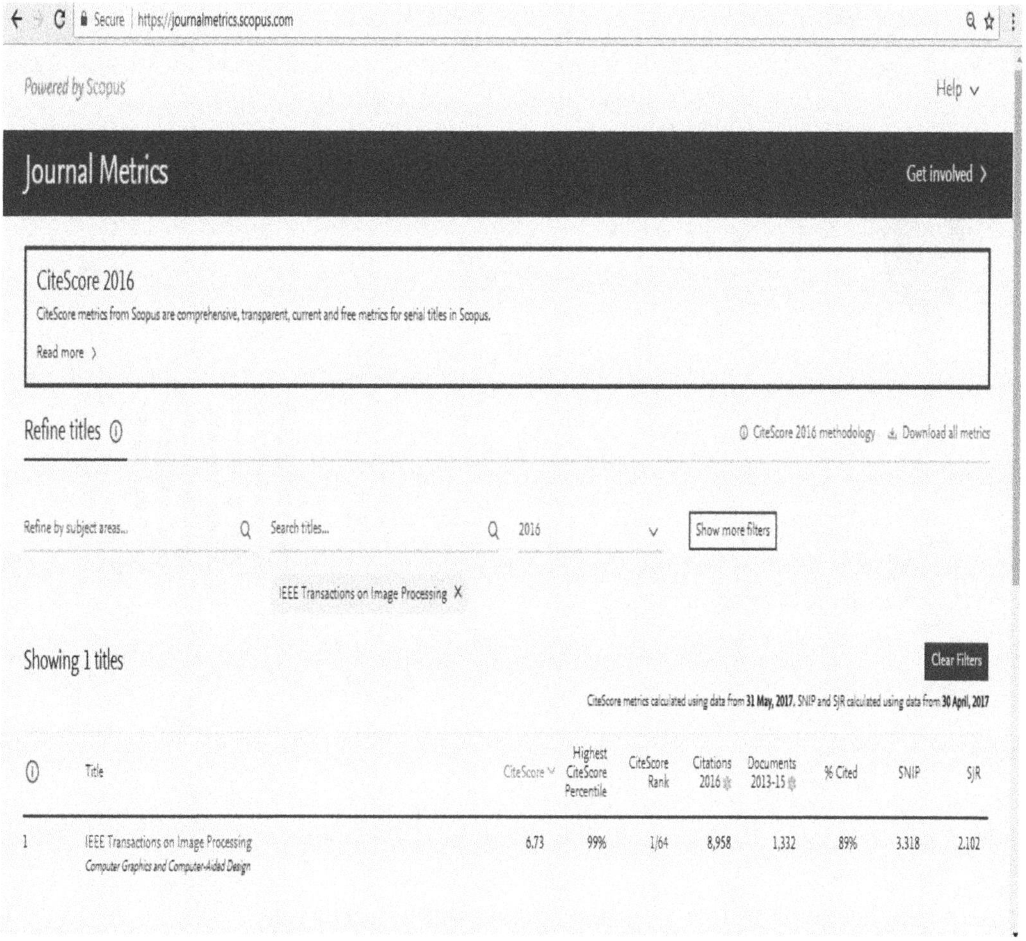

Presenting and Publishing Research Findings 223

FIGURE 6.3
Cite Score, SNIP and SJR of IEEE Transaction on Image Processing

of the researcher papers that have each been referred by at least "n" times by different papers

Calculation of H-index: List is created based on author's all publications and amount of citation received for every paper. Arrange the list in decreasing order of the quantity of citations every paper has received as shown in Figure 6.4. The H-index of author is 31, means there are 31 papers of author for which minimum 31 citations are observed.

Computing a researcher's h-index has some unmistakable favourable circumstances. It gives some level of straightforwardness about the impact they have in the field. This makes it simple for nonspecialists to assess a scientist's commitment to the field. H-index has a few hindrances, h-scores are available to control through practices like self-reference.

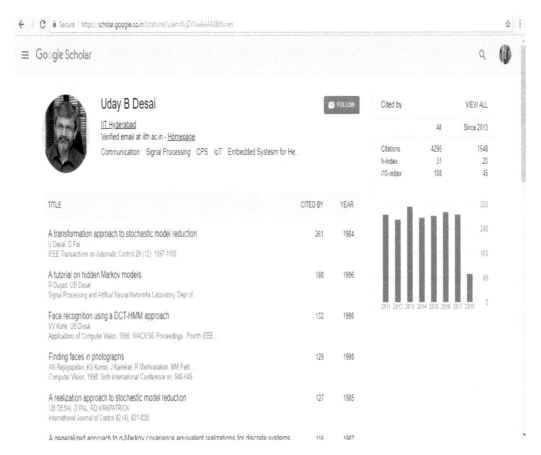

FIGURE 6.4
Google Scholar Citations of Dr. U. B. Desai, which Includes H-index and I10 index

- **I10-index** : The I10 index refers to the number of paper with 10 or more citations and is created by Google Scholar. The i10 index of researcher in Figure 6.4 is 108. This means that there are 108 papers of that researcher who have received citation more than 10 citations
- **G Index:** The G-index was proposed by Leo Egghe as an alternative to the H-Index.

    G-Index is calculated as: For a given a number of papers, which are ranked in decreasing order of the number of citations they have received, the G-Index is the maximum number, such that the top G articles received at least G^2 citations
- **Immediacy Index:** is a measure of the speed wherein content of specific journal is cited. The journal Immediacy Index gives the information about, how quickly articles in a journal are being cited [18]
- **Cited Half Life** is a measure of the "achievability" of how long content is referred after publication [19]

6.9.7 Research Gate (RG) Score

The RG score is measured by Research Gate. The RG score is the metric that measures scientific popularity based on how all of the work is received by peers. Authors RG score is primarily based on the published research and contributions to Research Gate obtained by peers. RG Journal impact is also provided by Research Gate.

6.10 Citation Index, Databases, Search Engines, Author Identifiers

6.10.1 Citations and Citation Index

- **Citations of researchers-** An accountable researcher must offer respective credit to different researchers and acknowledge their ideas. The citation should be appropriate as per respective publication standards of referencing
- **Self-citation:** Self Citation is citing own published paper during new publication. It just gives rise in overall citation count. Self-citation is not suggested until the previous works has strong relevance with the current work, and should be encouraged only when it keeps up the genuine necessity
- **Citation Index:** It is an index of citations of papers by a researcher. Google citation quickly shows the citation index but there are some drawbacks of Google citation experienced, as it will not cover journal that are not covered by Google scholar coverage and hence the dependency start with indexing. Citation score/count can be increase by focusing on simplicity and uniqueness of research keywords, step-by-step approach, articles, and results used in research paper/article [20]

  List of citation index and search engines are-
  - Shepard's Citations (1873): Eugene Garfield's Institute for Scientific Information (ISI) – ISI has introduced the first citation index for papers published in academic journals, currently known as Clarivate Analytics
  - Science Citation Index (SCI)
  - Social Sciences Citation Index (SSCI)
  - Arts and Humanities Citation Index (AHCI)
  - CiteSeer
  - IEEE Xplore
  - Scopus
  - Springer Link
  - Web of Science
  - PubMed
  - Google Scholar
  - Indian Citation Index
  - CNKI Scholar
  - Science Direct

Every indexing agency has their own criteria for journals getting indexed in their respective database. Each of above PubMed, Scopus, Web of Science, and Google Scholar have their own strength and weakness [21,22].

6.10.2 Most Preferred Citation Index and Search Engines

- **SCOPUS** is possessed by Elsevier. SCOPUS is a bibliographic database of peer-reviewed literature containing scientific journals, books and conference proceedings. It covers about 22,000 titles from more than 5,000 distributors, of which 20,000 are peer-assessed diaries in the logical, specialized, medicinal, and sociologies [23]. The journal indexed by SCOPUS assumed to be a reputed journal
- **Science Citation Index (SCI)** is a reference record initially delivered by the Institute for Scientific Information (ISI) and made by Eugene Garfield. It is presently owned by Clarivate Analytics (beforehand the Intellectual Property and Science business of Thomson Reuters). Science Citation Index covers in excess of 8,500 outstanding and critical journals, crosswise over 150 orders, from 1900 to the present. The journal indexed by SCI assumed to be of higher reputed journal [3]. The Science Citation Index (SCI) covers areas of life science, clinical solution, physical science, farming, science, veterinary prescription, building and specialized parts of incorporated recovery periodicals. Clarivate Analytics consider numerous variables while assessing journals for scope going from the subjective to the quantitative [24]

6.10.3 Some Other Indexing Services

- **EI (Engineering Index):** Dr. John Butler professor from Washington University founded Engineering Index in 1884. In 1998 Elsevier purchase EI. EI compendex has over 20 million index with more than 8, 00,000 added each year. EI has over 12 databases containing e-books, patent and abstract
- **ISTP (Index to Scientific & Technical Proceedings):** The scholarly article record of science and innovation was established in 1978. The ISTP index includes documents from life science, physics and chemistry science, agriculture, biology and environmental science, engineering, and applied science disciplines. The scholarly document can be a meeting documents or general conferences papers, forums, seminars, workshops, conference, and so on. Database consists of around 35% of engineering and applied science literature

6.10.4 Unique Author Identifier for Efficient Search

There is possibility of authors having same initial name and surname are working in same domain of research and they are publishing the papers in scholarly journals. It is very difficult to identify the exact authorship only from name of authors. Unique author identifies are used to deal with such situation of ambiguity in author identity.

Using such ID, research work by the same author in various journals and conferences can be searched uniquely, provided it is linked to authors.

Many distinctive author identifiers are developed, however widely unique author identifies are: SCOPUS Author ID (developed by Elsevier and employed in SCOPUS and connected products); ORCID (developed by ORCID opposition, that is nonprofit, community-driven organization, based mostly within the United States, but with international membership) and Researcher ID (provided by Thomson Reuters).

- **ORCID**: It is unique digital identifies given to author. You can register on-line on following site to induce ORCID ID [25] (Refer Figure 6.5 for registering at ORCID).

    Eg: Orchid Id of Dr Krishna Warhade is 0000–0002-0282–9766

- **SCOPUS ID:** SCOPUS identification ID is routinely assigned to all authors. The author can search papers by means of authors name or SCOPUS identity on the Web link [26]

    Eg: SCOPUS author ID of Dr. Krishna Warhade is 26656459200

- **Researchers ID:** Researcher ID is employed to spot authors and allows users to make a publication profile and generate citation metrics from internet of Science. Researcher ID provides a novel way to researchers to manage their publication lists, track their cited counts and h-index, establish potential collaborators and avoid author misidentification [27].

    Eg: Researcher ID of Dr. Krishna Warhade is E-5841–2018

In addition to these IDs, Google Scholar and Research Gate also provides author specific login with h-index and RG score metrics, respectively. Authors can open account in google scholar and ResearchGate to share their research profile to the world.

Google Scholar is a freely accessible internet search engine that indexes the full text or metadata of scholarly literature, while, **ResearchGate** is a social networking site for scientists and researchers to share their research, to raise queries and to propose solution to researcher's queries, and locate collaborators.

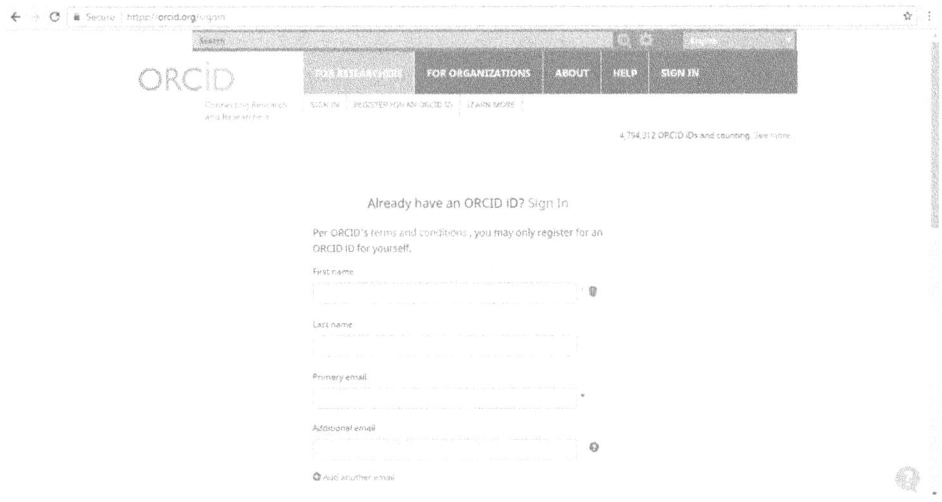

FIGURE 6.5
ORCID ID Registration and Sign In

6.10.5 Online Free Repository

There are various online free repository available like ResearchGate, academia.edu, Kidos,/own College website, and so on, where researchers can deposit their research papers. Such papers are freely available.

Guidelines to be followed while putting your papers in free repository like ResearchGate/own College website, academia.edu, Kidos, and so on.

- **Do's:** Putting research data/papers on repositories is always a good option, it will enable users to take out the data remotely without depending on his/her laptop or computer. There are various repositories available to trace out and reuse our research information/papers/article just in time. Such repositories are a quick way to share author's information to other researchers as well. Such practices shall increase researchers citations because research papers/contents are easily available to others researchers also

- **Don'ts:** However, there are some concerns, where researchers need to look upon. If, the journals and conferences are under copyright agreement, do not share paper directly with others, instead provide URL of the journal website where the papers are available. Check the copyright rules and regulations of the journal before putting your paper on repositories, otherwise it may consider as copyright infringement. You may maintain the privileges as public, private or protected (shared to limited users)

6.11 Use of Documentation Tools, Bibliography Tools, and Presentation Tools Useful for Writing and Presenting Paper and Theses

The term "bibliographic" means "A list of the written or existing sources of information on a particular subject." The "bibliographic database" is the database of bibliographic records organized in specific way. Here is a list of few bibliographic database tools, documentation tools, and tools for writing paper and theses,

- **Mendeley bibliographic tools**: Mendeley is a product of Elsevier. It is useful in managing the references. It consists of application accessible for Windows, MacOS and Linux. Mendeley requires the client to store all essential reference information on its servers with specific comments in database. Upon enrollment, Mendeley furnishes the client with 2 GB of free web storage, which is upgradeable on utilizing certain cost
- **Easy Bib:** Easy Bib is user-friendly citation option with an automated bibliography maker, and having three reference format alternatives: MLA, APA, and Chicago
- **Zotero:** Zotero is a free tool that helps the researchers to gather, cite, share, and organize research sources. It collects researcher's studies and reference documents in a single, searchable interface
- **RefDot:** This is an expansion for Google Chrome program. RefDot enables you to naturally bookmark and refer to any site or online article, and Amazon book pages

Presenting and Publishing Research Findings 229

- **Citelighter:** Citelighter provides a browser extension and allows researcher to bookmark the segments of website pages, spare them with notes, and make a list of sources
- **Citefast:** Citefast is a simple and free programmed reference index producer It allows the researchers to generate, organize and manage the citations quickly and accurately
- **EndNote:** It is product of Clarivate Analytics. Using endnote researcher can create and manage references from hundreds of online resources. One can read, review, and add comment in it. Researchers can create their own rules to automatically organize the references as per requirement. One can share his/her library to others
- **Bibme:** It's a definitive (free!) programmed book index producer with auto-fill. With Bibme, researcher should simply look for a book, site, article, or film, or fill in the data yourself. One can add to his/her own book index and download it in any required format of references like MLA, APA, Chicago, or Turabian design
- **Noodle Tools:** Noodle Tools is an online asset that gives you a chance to make references for any source. Researcher has to select the required style (MLA, APA, or Chicago)
- **RefWorks:** RefWorks is an online research management, writing and collaboration tool. It helps researches to easily gather, manage, store and share all types of information. It also helps them to generate citations and bibliographies
- **Latex**: LaTeX is used to write reports, theses, conferences, and journal papers. LaTeX is considered as standard freeware for the publication of scientific documents
- **Grammarly**: It is an online spelling and grammar checker to check your documents. There are many more online Grammar checking software freely available
- **Typeset:** It allows you to auto format your paper in any journal template [28]. Refer the screen shot Figure 6.6 for sign in and the features of Typeset
- **eXtyles:** eXtyles is a software product from Inera. It is a tool, which allow publishers to convert word file from any sources into required standard style document including citation style. It thus avoids time-consuming process of writing paper and provides error free document. It is an editorial and XML tool, which is preferably used by publishing staff of many publishers. Refer Figure 6.7 for details of the tool [29]

One can also look for other tools like colwitz, NinjaEssays, ReadCube, Qiqqa, and Citavi, which will help you to simplify writing of your research work.

6.12 Presenting Research Papers, Reports and Theses

Presentation of the research is very important and leads to discussion and better understanding of problem statement to examiners and other researchers working in the same area.

Following points are important while presenting your research to research community.

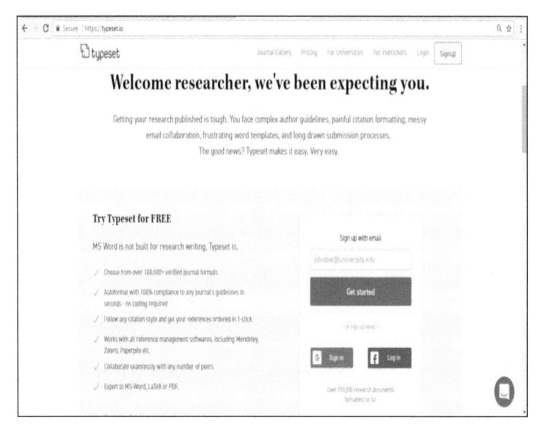

FIGURE 6.6
Sign In for Typeset

- The presentation contents should be simple, clear, and audience-knowledge-specific. Practice presenting it for improvement of presentation skill and confidence
- Distribute handouts of the presentation to examiners and other researchers prior to the presentation
- Good time management of presentation is a key of impressing the audience. For paper presentation, generally 15–20 minutes are given, whereas for theses presentation 45 min to 90 min are provided. Timing may vary and depends on conferences and Universities guidelines. Prepare number of slides according to the time allotted to you. The presenter should be ready with any dynamic change in presentation such as change in timing and duration
- Start your presentation with confidence. Be calm and relax. Maintain eye contacts with audience and talk loudly and clearly
- Focus on the proposed methods and experimental results. Convey your contribution in the research through tables and figures
- If the questions are raised during presentation, understand the question and give meaningful answers with facts and proofs

FIGURE 6.7
eXtyles from Inera for Publishers.

6.12.1 Preparation and Presentation

Novel ideas are recognized with its effective demonstration. The presentation should be clear, confident and sufficiently practiced.

(i) Audience Focus:
- Presentation should be delivered in such a way that it should relate to other presentation, keynote sessions and theme of conference
- Ensure correct choice of words to retain the curiosity and interest of the audience
- Eye contact should be maintained with audience throughout the presentation
- Maintain audience awake by adding cartoons, animations and videos in the slides and make sure the visual aids and audios are audible
- Maintain backup of slides for transparencies

(ii) Explanation:
- Many ideas should not be highlighted on the same concept
- Only five minutes should be spent for explaining the importance of the work

- Five minutes should be spent for key idea as an approach for the solution
- Five minutes should be spent for methodology and algorithm/flowchart
- Five minutes should be utilised for comparing the results and prove it better than the existing solutions
- Rest three minutes should be used for conclusion and future work

(iii) Tips for Slides:

- Keep only few slides excluding your Introduction and Question/Suggestion slides
- A slide should contain abstract, agenda, existing system/proposal, methodology, algorithm/flowchart, two slides for experimental results, comparison, conclusion, and reference
- Include color slides only for results and use professional templates
- For text use large fonts and do not put more text on slides, do not make audience make busy reading the slides instead of listening
- Slides in the text should be grammatically correct
- Avoid using gratuitous animation
- Use illustrations to explain complex algorithms
- Omit minor details, focus on the important concepts
- Presenter may not have enough time to discuss all ideas clearly
- Better to explain one idea in detail, other ideas briefly
- Anticipate technical questions, and prepare explanatory slides
- Accent may not be easy to understand
- Talk slowly in an audible range with modulation in voice

6.12.2 Presenting Posters

Poster presentation is included in most of the conferences and also used in academics for project presentation. It provides good opportunity to present visual summary of the research and innovation to the larger audience. Poster should be self-explanatory and academically sound. Poster presentation is a way to share your findings and results with the researchers working in your area. Poster is a visual aid for effective communication with conference delegates. Poster presentation success depends on preparation of an effective poster. Effective Poster presentation can improve your reputation in your organization and outside your organization. It is a way to disseminate research findings; hence they must be prepared and presented appropriately.

Quality poster can result in starting collaboration, fetching funds for the project and developing researcher's network. There will be dedicated time allotted in conference to see posters and discuss the work with delegates. Be present during that time and give attention to all delegates. You will be given a set of guidelines from the conference to prepare poster presentation. As there is a limitation of space in poster, text must contain sound scientific content with clarity. For developing a quality poster, guidelines and sample templates are available on various web sites. Select font properly which is easy to read. Do not use more than two fonts, use limited colors, and use bold, italic, or underline to show important words.

Posters typically include following sections:

- **Title:** Title is the most important part of poster, as it draws attention of delegates. Title should be short but long enough to indicate the nature of the study. It should be seen from a distance of 5–10 feet. Capitalize the first letter of each word
- **Authors:** Put the names of all authors and their affiliations below the title. If you are working on a collaborative project, then include name of collaborators and their affiliations
- **Introduction:** In introduction, state the importance of the topic, literature about the topic and what is already achieved by others and brief about your work
- **Methods:** Mention the algorithms or procedures use to study. Also elaborate about the primary or secondary data used and how the data is analyzed
- **Results:** the analysis and findings of the research must be illustrated by graphs, charts, table or pictures. Preferably by using graphs, this may save space in poster
- **Conclusion:** State clearly your research finding, its importance and future scope
- **References:** Keep the list of references short in poster, preferably three to five important references can be included
- **Acknowledgment:** Acknowledge the people who have contributed in your research, including your advisor. If you are working with collaborators, their names should also be acknowledged. Thanks the funding agencies for funding your project

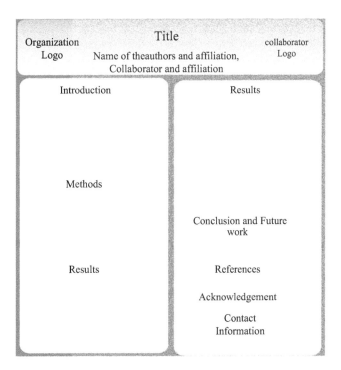

FIGURE 6.8
Template for Poster Presentation

- **Contact Information:** Include your full name, official address, and email. Provide URL of your own home page or the department home page. This will help delegates to contact you in near future for collaborative or consultancy work. Refer Figure 6.8 for poster presentation template

6.13 Summary

Presentation and publication is vital aspect of research. There are various publications options are available to author. Publication in reputed journal is credit-full to authors. Journal reputation is determined by journal impact in the field of research. Journal metric provides an extra insight into the different aspects of the journal's impact. It also helps authors in selecting a journal when submitting an article for publications

Further Reading

1. Chittaranjan Andrade, "How to Write a Good Abstract for a Scientific Paper or Conference Presentation," Indian J. Psychiatry, vol. 53, no. 2, pp. 172–175, Apr–June 2011.
2. Justus Randolph, "A Guide to Writing the Dissertation Literature Review," Prac. Assess. Res. Eval., vol. 14, no. 13, pp. 1-13, June 2009.
3. http://mjl.clarivate.com/cgi-bin/jrnlst/jloptions.cgi?PC=K
4. http://www2.le.ac.uk/library/for/researchers/publish/choose_where_to_publish
5. http://www.lib.vt.edu/find/journals/select-journal-publish.html
6. http://cms.iopscience.org/0659756d-d278-11e1-9b7a-4d5160a0f0b4/choosing-submit-paper.html
7. http://www.scimagojr.com/journalrank.php
8. http://journalfinder.elsevier.com/
9. http://www.springer.com/?SGWID=0-102-12-988548-0
10. https://portal.issn.org/
11. https://library.iit.edu/find/journals/by-issn
12. http://www.worldcat.org/
13. http://nsl.niscair.res.in/ISSNPROCESS/issnassignedinfo.jsp
14. https://clarivate.com/essays/impact-factor/
15. http://ipscience-help.thomsonreuters.com/incitesLiveESI/8275-TRS.html
16. https://www.scijournal.org/
17. https://journalmetrics.scopus.com/
18. http://ipscience-help.thomsonreuters.com/incitesLiveJCR/glossaryAZgroup/g7/7751-TRS.html
19. http://ipscience-help.thomsonreuters.com/inCites2Live/indicatorsGroup/aboutHandbook/usingCitationIndicatorsWisely/citedHalfLife.html
20. https://www.aje.com/en/arc/10-easy-ways-increase-your-citation-count-checklist/
21. htts://www.researchgate.net/publication/5958226_Comparison_of_PubMed_Scopus_Web_of_Science_and_Google_Scholar_Strengths_and_weaknesses
22. https://en.wikipedia.org/wiki/List_of_academic_databases_and_search_engines
23. https://www.elsevier.com/solutions/scopus
24. https://clarivate.com/essays/journal-selection-process/

25. https://orcid.org/signin
26. https://www.scopus.com/freelookup/form/author.uri?zone=TopNavBar&origin=NO%20ORIGIN%20DEFINED
27. http://www.researcherid.com/Home.action
28. https://typeset.io/
29. https://www.inera.com/extyles-products/extyles

7
Plagiarism

Mousami V. Munot, Sesha S. Srinivasan, and Anand S. Bhosle

CONTENTS

7.1 Introduction	238
7.2 Plagiarism	239
7.3 Reasons for Plagiarism	240
7.4 Types of Plagiarism	241
7.4.1 Plagiarism of Words	241
7.4.2 Stitching Sources (Potluck Plagiarism)	241
7.4.3 Patchwork Plagiarism	241
7.4.4 Self-Plagiarism (the Self-Stealer)	242
7.4.5 Cyber and Digital Plagiarism	242
7.4.6 Accidental Plagiarism	242
7.4.7 Plagiarism of Authorship	242
7.4.8 Plagiarism of Ideas	243
7.4.9 Reuse of Programming Code	243
7.5 Plagiarism in Various Fields and Its Consequences	244
7.5.1 Music Industry	244
7.5.2 Plagiarism in Academia	245
7.5.3 Plagiarism in Journalism	247
7.6 Software Used for Identifying Plagiarism	247
7.7 Plagiarism Policies	252
7.7.1 IEEE (Institute of Electrical and Electronics Engineers)	252
7.7.2 Springer	252
7.7.3 Elsevier	253
7.7.4 Committee of Publishing Ethics (COPE)	254
7.7.5 Government Rules and Polices	254
7.8 Techniques to Avoid Plagiarism	254
7.8.1 Referencing	254
7.8.2 Paraphrasing	257
7.8.3 Tips and Strategies to Avoid Plagiarism	259
7.8.4 Creative Commons License	261
7.9 Summary	261
Further Reading	262

> If you steal from one author it's plagiarism; if you steal from many it's research.
> —Wilson Mizner

Learning objectives of this chapter are to:

1. Comprehend and adopt best practices that would ensure the writer to avoid plagiarism.
2. Understand the various policies of various reputed journals such as IEEE, Springer, and Elsevier and so on.
3. Apply various techniques of avoiding plagiarism to ensure ethical and integral writing
4. Utilize various free and paid software for identifying plagiarism and implement paraphrasing

This chapter will enable the researcher to:

1. Cite and paraphrase the document effectively
2. Create/submit/publish a plagiarism free document/manuscripts
3. Ensures the originality and uniqueness of the work.

7.1 Introduction

Intellectual morality that guarantees integrity is indispensable to avoid any scientific or academic misconduct. While presenting and publishing any document, it is imperative to ensure its uniqueness and credibility. Attempts of manipulation, fabrication or falsification of the data, experiment, process, equipment, procedures, observations, findings, or results lead to issues of misconduct and are extremely unethical. Plagiarism is another most common type of misconduct and, in some jurisdictions, it may also be illegal. While most of the people in general have just nominal degree of awareness about plagiarism, not everyone is precisely educated about it. The Josephson Institute Center for Youth Ethics surveyed 43,000 high school students in public and private schools and found that:

- 59% of high school students admitted cheating on a test during the last year; 34% self-reported doing it more than two times
- One out of three high school students admitted that they used the Internet to plagiarize an assignment

This report was published in June 2017 (https://www.plagiarism.org/article/plagiarism-facts-and-stats).This chapter discusses plagiarism at length and enables the readers to gain insight to what is plagiarism, related offences, ways of avoiding plagiarism, and the repercussions of plagiarizing.

7.2 Plagiarism

Plagiarism is an inappropriate act of claiming credit or ownership of someone else's ideas or unfair annexation of another's work and entitling it as your own. Such actions are very common since the existence of human life on the universe. For as long as there have been poems and poetics, music and musicians, science and scientists, math and mathematicians, and so on, time has witnessed acts of seemingly unnamed literary thefts in every possible domain/field and by every possible means.

The origin of word "plagiary" (a derivative of plagiarus,) dates back to the first century AD and it involves a Roman poet, Martial, who complained that another poet had "kidnapped his/her verses." Although the word "plagiarism," took 15 centuries for its first appearance in English (in the various battles among Shakespeare and his/her peers), the Oxford English Dictionary credits Ben Jonson as the pioneer to introduce and use it in print in 1601 to describe someone guilty of literary theft. Since then, both the act and the rise of importance of the word "plagiarism" exponentially increased and has put a very high value on "originality" in any contribution. All of the developed and developing countries across the globe now consider expression of original ideas as intellectual property, which is protected by copyright laws, just like original inventions.

According to the *Merriam-Webster's Online Dictionary*, plagiarism is an act of fraud that involves both **stealing** someone else's work and **lying** about it afterward. To "plagiarize" means to:

- Steal and pass off (the ideas or words of another) as one's own
- Using other work or idea (another's production) without crediting them
- Commit literary theft
- Present as new and original, an idea or product derived from an existing source

The copying words or sentences of the other work in our work without citing, crediting their work and claiming as own and failing to put in marks or giving incorrect information about the source of a quotation, changing words but copying the sentence structure of a source or copying so many words or ideas from a source that it makes up the majority of your work, whether you give credit or not, are all examples of plagiarism.

Other form of plagiarism include intentionally copying lengthy passages from a book or journal article, or purchasing or downloading whole papers and submitting them as your own work leading to misconduct.

Plagiarism, intentional or unintentional, is always seen as clear violation of the code of ethics and the ramifications, in any case, remain the same. Many times the words or the text copied is not "substantial"; however, the action is still reflected as plagiarism. Consider the following examples:

(i) A researcher working on a problem in biomedical imaging uses Delaunay Triangulation to identify and disentangle overlapping chromosomes and publishes his/her contribution in a reputed journal. Another researcher working in the domain of automation of manufacturing processes copies just one word, "Delaunay Triangulation," and publishes an algorithm/methodology for extrication of overlapping using gears. Only one word is plagiarized in this case, but it is the key word contributing to a novel approach and is reported in biomedical domain. Unacknowledged use of the same concept in a different domain of manufacturing process is still definitely considered as plagiarism of idea.

(ii) A postgraduate student working on a project copies the aim, objective, and hypothesis from another report, assuming that actual number of words that are copied are relatively less compared to the total number of words in the entire report and therefore the overall similarity report of the dissertation will be within limit. Such cases are undeniably categorized as plagiarism cases because the content copied, although is a small part, is very important to overall dissertation. In another scenario, consider a researcher writing his/her PhD theses. He/she attempts to use survey paper already published by another researcher as one of the chapter in his/her own theses. A literature survey is not the most critical factor in the theses and does not usually have novelty in it. It is usually a general summarization of reported studies/finding and issues, and therefore act of using a published survey paper as a chapter in the theses may apparently seem obvious. However, such an act is still upheld as plagiarism. The dictionary meaning of the word "substantial" may therefore not necessarily be always applicable from the perspective of "plagiarism." It depends on the significance of the text (images, photographs, tables, and so on) that is copied.

7.3 Reasons for Plagiarism

There is lot of research going on to analyse the reasons people choose to intentionally plagiarize or unintentionally end up cheating. Some of the most obvious reasons are listed here:

(i) Most popularly reported reason, people plagiarize is the human nature of procrastination. People tend to postpone the expected and inevitable tasks/assignments/exercises until the last minute. This leaves them very limited time before the deadline and therefore important issues like integrity of the content are paid meagre attention. Sometimes procrastinators recurrently and deliberately look for justification for adjourning the work and avoid it. It is a short-term panic response that in the long run reflects their perpetual struggle against themselves, their indiscipline, and sometimes their hidden fears, thwarting their career paths. With consistent efforts, it is possible to overcome the habit of procrastination and ensure the integrity and honesty in the assigned work and contributions.

(ii) People lack knowledge about proper citations. People end up casually using readily available information in the electronic documents or pages on the Internet and other sources without seeking their permission and acknowledging or citing the source. Response to loads of information available at their fingertips, which can be comfortably replicated/duplicated, also leads to plagiarism.

(iii) Students are not fully aware about plagiarism and they do not fully understand the conventions required in the academic/scientific/technical/non-technical writing.

(iv) At some instances, people are overconfident about their skill set and believe they can escape and get away with plagiarism. They are comfortable to plagiarize when they realize that the content was not submitted to plagiarism detection tools or not uploaded on other internet sources.

(v) Sometimes negative emotions and competitive urges to gain the credits for the module without sincerely working enforce conscious decision of plagiarizing on the students.

Colleges and universities are taking efforts to educate the society about plagiarism and have launched massive online open courses about plagiarism. Institutes have their own policies that define plagiarism and it is a standard practice to establish strategies for dealing with cases of plagiarism. The consequences of plagiarism are usually severe, ranging from an automatic "F" in the respective course to temporary suspension or even permanent expulsion from the institute/conference or journal. Some of other multitudinous penalties include loss of personal pride because of cheating yourself and letting yourself down. It ultimately leaves the plagiarist with fabricated levels of attainment, fulfilment, satisfaction, and glory on personal fronts.

7.4 Types of Plagiarism

There are various ways of categorizing the act of plagiarism. Some of the types are detailed later in this chapter.

7.4.1 Plagiarism of Words

This is also called as "The Ghost Writer." The writer copies every single word from the source without use of quotation mark, citation, acknowledgment, or attribution. Ghost writing or plagiarism of words involves direct use..........

(Direct use of the entire statement/definition makes the document being entitles as plagiarized and is infringement of academic policy or code of ethical conduct.

7.4.2 Stitching Sources (Potluck Plagiarism)

The internet is flooding with large amount of informative data from numerous sources that are readily available for use. Writer tries to fetch essential and appropriate content by copying from several different sources. Modifying and stitching the relevant contents from various documents, books, online sources, web pages and repositories to make them fit together and look appropriately linked leads to potluck plagiarism.

7.4.3 Patchwork Plagiarism

The third type called patchwork plagiarism. Such plagiarism occurs when we borrow the content from original source and weaves them into our paper without citing. Here sources are used more than one like above. Suppose a student obtain four sources from where he/she copied one sentence from A source, one sentence from B source, one sentence from C source and one sentence from D source, and so on. Here the student thought that they are not copying anything and even they cited the references of sources. But still this is plagiarism due to unorganized and not composed sentences they used in their work. In such a case, direct quoted words should be cited with quotations.

7.4.4 Self-Plagiarism (the Self-Stealer)

Self-plagiarism is the use of our own previous work in another new context without citing that it was used previously. The idea behind is the reader should know that this was not first use of the material. Therefore republished text work is considered as self-plagiarism and the credit received from previous work is considered as corrupt. Hence the writer must cite when using previous written work.

7.4.5 Cyber and Digital Plagiarism

Cyber digital content includes the information posted by authors on the Web such as e-books, notes, graphs, video, audio, music, images, online databases, questionnairies, electronic reserves and so on. Such information is usually protected using copyright. However, it is mistaken, as a public area property and is used by the people leading to cyberdigital plagiarism. The author needs to take the permission from such website and include the date when the material was accessed or downloaded.

7.4.6 Accidental Plagiarism

Accidental plagiarism occurs due to the ignorance of a person resulting in failure to properly paraphrase, quote and cite their research. Improper method of documentation results in misattributing. This means you are claiming someone else's work as own. In other words, if you have taken reference from a book or may be expressed the meaning of a paragraph of an article in your research work, but you do not include an in-text citation, then the reader will assume that the idea/words are yours, not someone else's. This is nothing but an example of authenticity violation. Only by mentioning the reference of the sources in reference list it is not enough. The missing citation at text used makes it plagiarism.

It is sole responsibility of author to understand when and how to cite the material used and provide references to the sources. If an author knowingly claims ownership of already reported work, it is considered as intentional plagiarism. It includes buying papers online, copying and pasting the information from any available sources. Accidently failing to cite the references appropriately is an un-acceptable excuse and such an attempt is still considered as plagiarism.

7.4.7 Plagiarism of Authorship

The authorship means presenting the other author work with simple modification of sentence and presenting as own work.

7.4.8 Plagiarism of Ideas

Using and presenting somebody else's idea as your own and taking away all the credit without any reference to the original idea and also submitting a paper by incorrectly citing another's ideas. This is where the student uses one of the following as the basis for the whole, or a substantial part, of the assignment:

- published or unpublished books, articles, reports, or magazines
- some resources from the Internet
- some sort of ideas pitched in a TV program or a radio program
- an essay from an essay bank
- a piece of work previously submitted or to be submitted by another student
- from any newspaper article
- from an unpublished manuscript or record book

7.4.9 Reuse of Programming Code

With exponential increase in the number of developers and available software, the reuse of the programming code is usually encouraged, but it is important to realize the fact that such programming projects are done to improve one's individual skills.

You learn very little when you copy the programming code from other source. Even you do not understand the logic of the code written by other. If you are using other people's code, to avoid being accused of plagiarism make sure to give proper references and citations. You must clearly state the source of the code, for example, name of author, page in the book that you have taken the code from, web page URL address. Not giving reference to the code you have used is a plagiarism offence and is dealt with as such.

Note you will always awarded good marks for different unique code written to solve the problem and the originality you used, than the code you use from others. If you don't do any code on your own and just copy the entire program (referencing the authors), you will be awarded no marks as you have made no contribution to your coursework.

There are two main detection techniques of plagiarism. One is manual and another is software-assisted. It's very difficult to check plagiarism by manually because it requires good and excellent skills and memory to identify plagiarism. Even it's very difficult for human being to check all documents on various millions of sources. Hence it is wiser to go for the software based testing for plagiarism, where the software automatically checks all sources and generates the report automatically as will be detailed further in this chapter.

The act of plagiarism comes from copying both published and unpublished work from different sources. It's not just copying from published work of source but includes unethical use of unpublished sources as well. Moreover, just copying the other work and representing it in your work by simply modification and arrangement of few words without proper citation at text place and without showing the direct copied sentence by quotation and references at reference list is still plagiarism. Such things are easily detected in the modern tools like Turnitin.

Plagiarism is an act of fraud that is stealing some persons work and then lying after is absolutely unacceptable. If you are under pressure of any workload you may get many ideas, you may take shortcut to do your work, you may copy the things from other work but the idea of thinking is bad, find the solution, find new ideas, solution to do the work. Don't feel under pressure, if you are finding it hard take advise from your experts ask them and try something new you will get complete new results don't worry if it's bad or good, don't think of future consequences. You never know what coming in your way it may be interesting and worthwhile reading. Even you'll get many ideas from the people appreciate them give credits they deserve.

7.5 Plagiarism in Various Fields and Its Consequences

Here we discuss few important domains where plagiarism act is considered seriously and have more impact on original work and violating copyright act.

7.5.1 Music Industry

Music industry generates revenue by selling the original copyright write songs/albums, even by live music events/shows. The music composition is very highly technical and creative task for composition of songs. It takes several months to compose good songs. Its highly hard work to create good composition of songs, but some people are desire to make money by copying other work that's why even the music industry also fall prey to plagiarism. According to music plagiarism law, the sound recording and music composition are included in separate copyrights. Both sound recording and music composition are considered separate work. Even we can copyright our published and unpublished musical work.

Let's take few national and international examples of songs.

In some situations the music plagiarism is very tricky act, some song may be accidently or not, they inevitably like other songs. It takes very difficult process to check music plagiarism, it can be neglected, when the similarity is less, but the things become messier when the similarity is more, in that case the copyright law get involved.

Few example international case of music plagiarism listed here.

- Miley Cyrus, "We Can't Stop" and Flourgon, "We Run Things"
- Ed Sheeran, Faith Hill, and Tim McGraw, "The Rest of Our Life" and Jasmine Rae, "When I Found You"
- Meghan Trainor's "All About That Bass" and Koyote's "Happy Mode"
- The Beatles, "Come Together" and Chuck Berry, "You Can't Catch Me"
- Jennifer Lopez, "On the Floor" and Kaoma, "Lambada"

Case 1: Miley Cyrus, "We Can't Stop" and Flourgon, "We Run Things"
Recently in March 2018 Jamaican reggae musician Michael May performed in name of Flourgon, but its filed a lawsuit against Flourgon claiming that the "We Run Things" song is copied and pulls from a 1988 Miley Cyrus song "We Can't Stop." According to the lawyers, the melody/rhythm/cadence/inflection, are similar to old Miley Cyrus song name "We Can't Stop".

Case 2: Robin Thicke and Pharrell Williams sued Marvin Gaye's family because the family claim that "Blurred Lines" violated the copyright of Marvin Gaye's song. The claim was filed in August 2013. After five years the lawsuit was finally settled in 2018. The jury found Robin Thicke and Pharrell Williams guilty and ordered them to pay $4 million in damage of violating law of copyright. They agreed to pay some profit of $3.2 million that they earned from song. This is one of the biggest judgments concerning copyright protections in the music industry.

Let's take list few national examples of songs.

- Teraa mujhse hai pahle kaa naataa koii (Aa Gale Lag Ja, 1973) and "The Yellow Rose of Texas" (Traditional) both were criticized by many people for violating copyright
- Mil gayaa ham ko saathii (Hum Kisi Se Kum Nahin, 1977) and "Mamma Mia" (ABBA, 1975) both were criticized by many people for violating copyright
- Dilbar mere kabtakmujhe (SattaPeSatta, 1982) and "Zigeunerjunge" (Alexandra, 1967) were both criticized by many people for violating copyright
- Tum se milke (Parinda, 1989) and "When I Need You" (Leo Sayer, 1977) both were criticized by many people for violating copyright

Let's move to the next section, which will cover the plagiarism in academic, how it beneficial for the student to protect their data from unauthorized use.

7.5.2 Plagiarism in Academia

Plagiarism also plays a vital role in academic for violating the student's work. It is considered as a biggest crime in academic. This is because students are always encouraged to think on unique idea. They put full efforts on their unique idea; hence, it is good academic practice for every university to care about plagiarism.

(i) Reputed institutions encourages student/faculty for writing original work.

The most challenging task for every good scholar is to take what is already done before/said before. And further use this material to propose his/her work in original ways.

Some student may ask you why we should unnecessary efforts be taken in adding redundancy to existing work or paper "if this research paper is already argued, why are we discussing the same topic?" The answer is as follows:

- It provides a model for our own work by doing research on the subject that you are interested and you become expert and familiar with existing updated works on the interested subject in day to day with technology
- It is valuable because, the presentation and discussion on the represented information is totally unique from the other
- It helps to add more information and new dimension into your subjects, which may be made available for all
- It is also an opportunity to present your findings, report or argument

Writing in academic may seem to be contradictory.

On the one hand, we ask you to:	But also to:
Always find what is written on the selected search topic and report it and demonstrating the things that written in the topic	Writing a topic in a original way.
Always try to bring in opinions of expert panel and respective authorities in frequently for betterment.	Keep on working and doing more than simply reporting them; don't accept the opinion directly try to remark on this opinion, include them, have the same opinion or oppose with them.
Notice articulate phrasing and learn from it, especially if you are trying to enhance your capability in English.	Make sure you are using your own words, when you putting these words into a paper directly.

It is very challenging to build an original framework and still acknowledge every idea/ reference used in your work. It is manadatory to preserve educational integrity and novelty of your honest work as well.

(ii) **Oxford university plagiarism rules.**

By employing the good academic principle and practices, it is very easy to avoid the plagiarism by adopting university mentioned rules. The university detail:

- Forms of plagiarism:
    - exact (word for word) quotation without clear acknowledgment
    - cutting and pasting from the Internet without clear acknowledgment
    - paraphrase
    - complicity
    - incorrect references
    - malfunction to acknowledge assistance
    - use of material written by professional agencies or other persons
    - auto-plagiarism
- Why does plagiarism matter?
    - A principle of honesty: all academic community should acknowledge their originator of idea, words and data
    - It is unethical: can impact serious problem on future career
    - It's under estimate the whole academic record
- Why should you avoid plagiarism?

There are many reasons to avoid plagiarism.
  - To impact and produce the highest quality work done by you
  - Improve your lucidity and quality of writing skill
  - Most importantly it provides honesty credibility and authority to your work

- What happens if you are thought to have plagiarized?
  - University put penalties of reduction in marks
- Does this mean that I shouldn't use the work of other authors?
  - Give the credit to the main author form where you got the idea and observation you cite
- Does every statement in my essay have to be backed up with references?
  - No it's not necessary to give references for fact that are common in discipline but before it we have to confirm whether it's common or not
  - But is safer side better to be citing the reference

Every University has their own policy against plagiarism.

7.5.3 Plagiarism in Journalism

Plagiarism is exponentially increasing in the field of journalism. The main assets of news organization are public accountability, professionalism, and reliability. The organization should report honest and accurate news when they report a story, from this the organization maintain reputation and ethical journalistic standards. Plagiarism in journalism affects their own individual reputation as well as harms organization who they represent

There are three essential types of plagiarism that occur in journalism.

- Collecting information from one reporter and passing it on media as their own news without giving credit to original reporter
- Copying others work and passing it on to newspaper as their own article
- Plagiarizing other's ideas as one's own

The plagiarism in journalism is rising rapidly. Recently the plagiarism was discovered in four stories. An online editor of WQXR resigned after 10 stories they filed are found plagiarized material. Recently many reporters are lost their job due to false filed stories. In 2012 CNN and Time suspended CNN's Fareed Zakaria when he found in plagiarized one of his/her *Time* magazine columns.

There are many more such cases of every possible way in absolutely every field and the consequences are much severe than mere financial. They extend to causes of defame, loss of integrity, respect and confidence at personal and social level.

7.6 Software Used for Identifying Plagiarism

After writing any paper or story, it's important to check the paper is free from plagiarism or not. Plagiarism free paper is very beneficial for representing quality of work. The testing of plagiarism is done by software. There are many software tools are used to testing plagiarism. The software available online is of two types

- Paid software (iThenticate, Turnitin, Urkund)
- Free software

(i) iThenticate

iThenticate is very famous software widely used in famous publications IEEE, Elsevier and Springer and so on. iThenticate also provide different file type that we can upload easily without any restriction of file type. The file types that supported to upload in iThenticate are Microsoft Word® (DOC and DOCX), Word XML, Plain Text (TXT), Adobe PostScript®, Portable Document Format (PDF), HTML, Corel WordPerfect® (WPD) and Rich Text Format (RTF). Even submitting paper may not exceed 400 pages and file size may not exceed 100 MB.

The features provided by iThenticate are:

- Advanced Plagiarism checking
- Match report
- Content tracking
- Review report
- Major matches
- Online sustain and guidance
- Multiple user collaboration
- API integration for organizations and many more

Example: Summary Report generated by the iThenticate tools.
The snapshot depicted in Figure 7.1 indicates the overall similarity breakdown report generated by iThenticate database. The generated report mode shows different color line of the matches found followed by the paper with the matching areas highlighted in different color.

(ii) Turnitin

Turnitin is widely used commercial internet based plagiarism software tool for plagiarism detection services launched in 1997. University and many more IPR filling organizations buy this tool for plagiarism detection. Turnitin checks the submitted document in database and gives the similarities with existing sources. Consider a manuscript submitted to Turnitin software for plagiarism detection. After submitting the paper, Software will check the manuscript line by line for similarity. If the content in the paper is found similar to the one already existing in database, it is highlighted using various colours. Same colours are used to match the copied sentence with its source for easy understanding and further in-depth analysis, if required. Figure 7.2 depicts a snapshot of Turnitin software indicating the similarity report of a manuscript submitted for plagiarism check.

The above report is marked with various colors, it indicates the similarity found in that particular colored line. Turnitin software, thus also calculates percentage of similarity found different paper that matched.

The final originality report is shown below in Figure 7.3.

From the originality report, the research can easily analysis the sources for possible similarities: 8% similarity index found in all paper, 3% content from internet source, 7% similarity found in different publication and 1% similarity found in student paper. The report shows student plagiarism of 1%. The material entitled under "student plagiarism" is unpublished. However, it is mandatory to take a serious note of even unpublished material to ensure integrity of our write-up. It is also important to take note of a possibility where the material entitled student plagiarism is actually your own write-up submitted earlier for

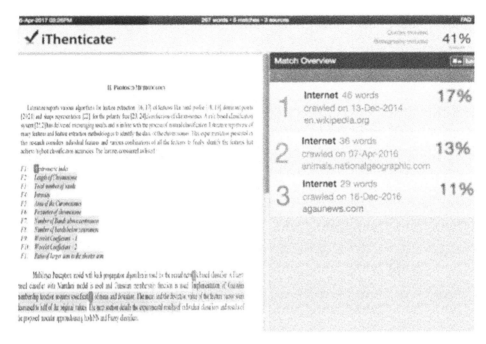

FIGURE 7.1
Similarity report generated by ithenticate (courtesy Turnitin)

Mousami V. Munot
Associate Prof. Dept of E&TC, Pune Institute of Computer Technology, Pune, India, munot.mousami@gmail.com

Abstract: The manual process of ordering of chromosomes (karyotyping) is long, tiresome and prone to errors. Automated classification of chromosomes in one of the appropriate 24 classes has therefore emerged as a thrust area of research. Researchers have studied the important features of the chromosomes which play a significant role in karyotyping of chromosomes and reports success of neural networks as promising classifiers. Perceptive vagueness in the texture of the chromosomes, however, encourages use of fuzzy classifier for efficient classification. This research initially revalidates the performance of neural network and fuzzy system individually, to examine the dominant features for each and every Denver group and sub-group and further presents a two-stage classification process to resemble manual approach of generating a karyogram. The proposed novel and intelligent approach initially identifies the best features for every class / subgroup and further utilizes them to build a modular Neuro-Fuzzy classifier for organizing the chromosome as per a standard karyogram. The proposed algorithm is validated using a publically available database comprising of 90 metaphase images (4320 segmented chromosomes). The results conforms the efficacy of the algorithm by achieving improvement by 2 % in overall performance of the proposed hybrid classifier and thus offers a practically deployable solution to the issue of manual karyotyping.

Keywords: karyotyping, chromosome classification, neural networks, fuzzy systems

1. INTRODUCTION

Chromosomes are the genetic carriers of the human body [1, 2]. Karyotyping is the procedure of envisaging and sorting the chromosomes within a cell during the metaphase stage of cell division [3, 4]. It is an indispensable assignment in cytogenetics for the diagnosis of genetic disorders [5]. Manual process of classification of chromosomes is tedious and prone to human errors [6]. It still faces lot of challenges and therefore is a long standing issue in the domain of medical image processing [7]. The task of generating the karyogram can actually be envisioned as a pattern recognition problem demanding recognition of the features and further classification of chromosome in its respective classes(24 classes with seven main groups called as Denver groups: A-G and 24 subgroups) [8, 9]. Literature describes overall prevailing features of the chromosomes considering the entire classification problem. Various approaches based on Hidden Markov Model (HMM)[18], pairing of chromosomes [9],

FIGURE 7.2
Result of submitted paper in Turnitin (courtesy Turnitin)

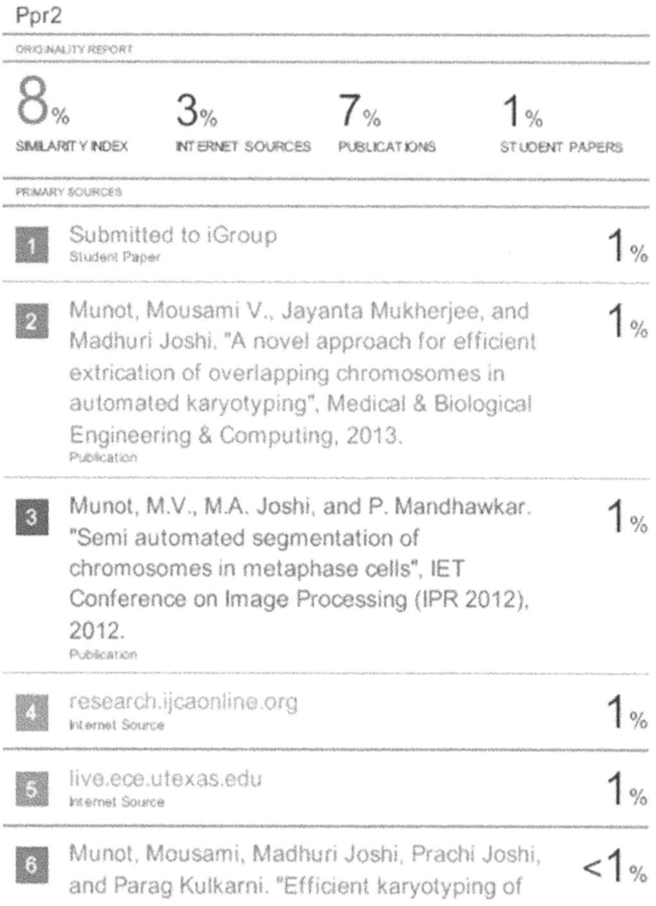
Ppr2

ORIGINALITY REPORT

8%	3%	7%	1%
SIMILARITY INDEX	INTERNET SOURCES	PUBLICATIONS	STUDENT PAPERS

PRIMARY SOURCES

1	Submitted to iGroup Student Paper	1%
2	Munot, Mousami V., Jayanta Mukherjee, and Madhuri Joshi. "A novel approach for efficient extrication of overlapping chromosomes in automated karyotyping", Medical & Biological Engineering & Computing, 2013. Publication	1%
3	Munot, M.V., M.A. Joshi, and P. Mandhawkar. "Semi automated segmentation of chromosomes in metaphase cells", IET Conference on Image Processing (IPR 2012), 2012. Publication	1%
4	research.ijcaonline.org Internet Source	1%
5	live.ece.utexas.edu Internet Source	1%
6	Munot, Mousami, Madhuri Joshi, Prachi Joshi, and Parag Kulkarni. "Efficient karyotyping of	<1%

FIGURE 7.3
Originality report generated by turnitin (courtesy: Turnitin)

plagiarism check. Turnitin software provides an option of saving/not saving the material to the repository thus giving the flexibility to the users. However, Urkund software by defaults saves the material into its repository without offering any option to its users.

(iii) Urkund

Urkund plagiarism software is widely used in throughout the Europe for plagiarism check. This software is completely automated used at universities and university colleges. Its check the submitted document against three central source areas:

- Internet
- Published material
- Before submitted student material

Based on this central search the Urkund software defines the percentage of plagiarism in the submitted paper and generate the report as depicted in Figure 7.4.

Plagiarism

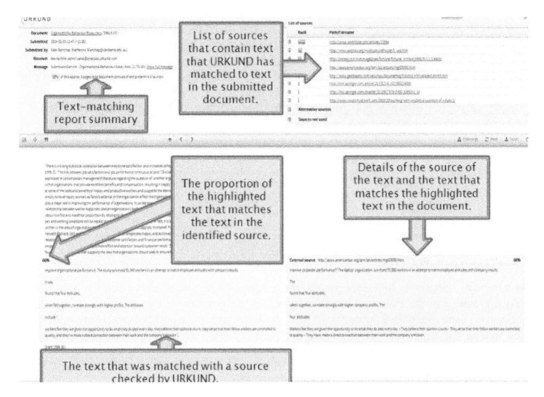

FIGURE 7.4
Urkund Summery report (courtesy Urkund)

(iv) Free software:
Modern technology and the Internet given us to access tons of information on the cloud of the Internet; hence everyone tries to access these information and extract useful information in their own work that related to research, study and so on.

To avoid the copyright and plagiarism there are many more free software tools available on internet. The most popular tools that are used widely for short plagiarism check included:

- Dupli Checker
- Copyleaks
- PaperRater
- Plagiarisma
- Plagiarism Checker
- Plagium
- PlagScan
- PlagTracker
- Quetext
- Viper

Dupli Checker is one of the most effective free plagiarism software detection tools on the Internet. It is free of charge, is user friendly and permits 50 scans per day for every registered user. However, for every unregistered user only one search per day is permitted.

All the freely available free software tools have their own rules and regulations and even they have few restrictions about the plagiarism check. Such tools are widely preferred in plagiarism check in small article with few 100 words.

Most of the software tools developed for identifying the plagiarism in material/document and manuscript are restricted to only checking and validation of the text content and the similarity report is generated accordingly. However, it is also necessary to examine the plagiarism in the diagrams, images and figures to ensure full integrity.

7.7 Plagiarism Policies

As already discussed, the documents/manuscripts submitted to various journals are scrupulously examined for plagiarism to ensure originality in the contents. Plagiarism polices of some journals (global level) and the Government (national level) are summarized in this section.

7.7.1 IEEE (Institute of Electrical and Electronics Engineers)

The IEEE (Institute of Electrical and Electronic Engineers) organization is global and professional body for working towards research, development and enhancement of latest technology in wide areas.

IEEE refuses plagiarism of prior processes, results, or words without acknowledging the proper sources. Various Engineers and Scientists from all around the world send their papers to IEEE for publishing. These papers are passed through rigorous plagiarism check and the papers that fail the test are rejected. Plagiarism at any stage is unacceptable. The levels of misconduct described in the sections are [26]:

- Level 1: 50–100% copied. (Uncredited Verbatim Copying of a Major Portion)
- Level 2: 20–50% copied. (Uncredited Verbatim Copying of a Large Portion)
- Level 3: < 20% copied (Uncredited Verbatim Copying of Individual Elements like Paragraph, Sentence, Illustration, and so on)
- Level 4: Improper paraphrasing
- Level 5: Credited but without clear delineation

7.7.2 Springer

Springer is a member of cross check. Springer also has online instrument (iThenticate) to check a composed conflict with that database. Springer is putting forth this screening programming to journal editors of Springer diaries and society and publishing partners diaries. A software can be utilized as a part of the publication procedure to recognize contents however, a manual examination of the coordinating content is yet required for judgment to recognize if copyright infringement has happened or not. The article "Publishing Ethics for Journals" (https://www.springer.com/gp/authors-editors/journal-author/journal-author-helpdesk/publishing-ethics/14214) on the Springer site tosses some light on the stand taken by Springer on the issue of unoriginality. It is anyway Springer's view that in remarkable cases (e.g., in instances of rehash guilty parties or creators utilizing injurious dialect) the Editor-in-Chief/Editorial Board of a Springer journal reserves the right and has the privilege to decline the manuscripts in cases of severe infringement.

7.7.3 Elsevier

Elsevier is also a member of cross-check. Elsevier follows some code of ethics which includes but is not limited to:

- Authorship of the paper: Three basic criteria must collectively be met to be credited as an author:
    - Significant input to the study conception and design, data attainment, examination, and understanding
    - Draft or revise the article for scholar content

    Approval of the final version
- Originality and plagiarism: The writers should ensure that they have composed completely unique works, and if the writers have utilized the work and additionally expressions of others, this has been fittingly referred to or cited
- Data access and preservation: The creators might be requested to furnish the crude information regarding a paper for article survey, and ought to be set up to give free to such information
- Acknowledgement of sources: Proper acknowledgment
- Disclosure and conflicts of interest: All entries must incorporate revelation of all connections that could be seen as displaying a potential irreconcilable situation
- Fundamental errors in published works: At the point when a creator finds a huge mistake or error in his/her own particular distributed work, it is the creator's commitment to instantly tell the diary supervisor or distributor and coordinate with the proofreader to withdraw or remedy the paper
- Reporting standards: Creators of reports of unique research should show a precise record of the work executed and in addition a target exchange of its importance
- Hazards and human or animal subjects: Articulations of consistence are required if the work includes synthetic substances, systems or hardware that have any unordinary risks intrinsic in their utilization, or in the event that it includes the utilization of animal or human subjects
- Use of patient images or case details: Studies on patients or volunteers require morals board endorsement and educated assent, which ought to be recorded in the paper

There is a plagiarism factsheet prepared by Elsevier that lists various unethical actions and ways to avoid them or prevent them.

7.7.4 Committee of Publishing Ethics (COPE)

The first meeting of COPE took place in April 1997 and it was a generally casual social occasion of a gathering of concerned editors. At that first gathering it turned out to be certain that the gathering had witnessed and experiences cases of research misconduct, fraud, unethical behaviour and misinterpretations. COPE is formulated at a global level to reinforce to importance ethics and integrity in publishing. It involves numerous distributors, all those associated with production and publishers as members. COPE has enriched resources and practices to assist the editors of scholarly journals, publishers/owners and educational institutions to preserve and promote integrity.

7.7.5 Government Rules and Polices

Government of all the developed, developing and even the under developed countries are striving to ensure integrity in academia and research. For example, in India, Ministry of Human Resource Development (MHRD) has taken strict actions leading to dismissal against the cases of plagiarism. With a view to check plagiarism in higher education, University Grant Commission (UGC) in India has notified the Promotion of Academic Integrity and Prevention of Plagiarism in Higher Educational Institutions Regulations, 2018. The objective of the regulations is to stimulate academic research and restrain plagiarism by developing systems to detect it.

7.8 Techniques to Avoid Plagiarism

One of the essential components that need to be inevitably cultivated and inculcated in human beings is a practice to take pride in your own work and enjoy genuine satisfaction in your own work with real level of attainment matching your actual ability. Beyond being unethical, unfair, or dishonest, Plagiarism of any kind, whether caught or not, is always cheating yourself and letting yourself down and should therefore be avoided.

7.8.1 Referencing

Key to avoid plagiarism is to acknowledge the sources used, whether published or unpublished.

Giving credit to the work of other writers or contributors in the corresponding domain exhibits highest level of integrity. Moreover, it showcases your efforts of rigorous ground work, thorough analysis and further also guides the readers with all available relevant sources, used by the author. Most importantly, when the material is aptly cited, it enables the readers and other likeminded community to clearly distinguish your ideas from the reported literature and appreciate the originality of your work.

While preparing your work, it is a standard practice to refer to records, volumes, articles, reports, newsletters, newspapers or other programs and documents. An article or a manuscript may therefore include wide assortments of statements, facts, opinions, results, findings, discussions and concluding remarks. Referencing the contents appropriately reduces the chances of suspicion of plagiarism; however, it may not be required to reference every sentence in the article. In order to aptly cite the references it is essential to understand the difference between statements, facts, and opinions.

Consider the following examples:

"India celebrates its independence day on 15th August."

This statement is neither a direct quotation nor is contentious. It is actually a very well-known fact and absolutely every Indian is aware of it. Such a statement in your assignment or document would not need any referencing or citing. While some facts are categorized as "common knowledge," it is important to note that lesser known facts may need referencing. There exists absolutely no globally acceptable agreement on the definition of "common knowledge." For example, consider the statement:

"Jawaharlal Nehru, the first Prime Minister of independent India, delivered a speech on 'Tryst with Destiny' on 14/08/1947."

It is considered as a common knowledge that Jawaharlal Nehru is the Prime Minister of free India, but the fact that he delivered a speech on *"Tryst with Destiny"* may not necessarily be equally well known and may not be unanimously considered as common knowledge. In all uncertain and doubtful cases, it is always strongly recommended to cite the source. (The book *Midnight's Children* by Salman Rushdie, the novel *Train to Pakistan* by Khushwant Singh.)

If the author further plans to include various statements of the speech delivered by all the leaders on 15/08/1947, quotation marks are indispensable. It is also obligatory to provide references and cite the sources. In text referencing is preferred when some slogans in the speech are included in the documents, whilst if couple of lines of the speech are copied in the write-up, it is necessary to use an indent (move the paragraph/lines away from the set margin of the document) so as to illustrate clear mark and extend of quotation. If the author finally concludes the assignment with his/her opinion (original and unreported) and unique concluding remarks, referencing is definitely not expected and in fact may contradict the originality or the novelty component of the assignment. However, if the opinion (positive, negative or neutral) is formulated based on some survey or existing reports, it is necessary to give due credit to the sources based on which your opinion is articulated. Thus it is imperative to understand and distinguish between facts, lesser known facts and opinions during the process of referencing.

Reference list and bibliography are often used interchangeably, which may or may not be acceptable to journals/newsletters depending on their policies. To ensure the transparency in our presentation, it is also necessary to understand the difference in citation and referencing. As quoted by Dr. Sarah Eaton, an educational leader, researcher, author and professional speaker, in her blog "Knowing the difference between a citation and a reference is one of those subtle details that move you from the category of *"novice researcher"* to *"respected researcher ."*

These terms are explained here:

(i) Citation: It is an act or practice of mentioning/attributing/indicating a specific source or quote material in the body of your paper. MLA, Vancouver referencing styles, and American Psychological Association (APA) are common styles or formats of citation. Nevertheless any citation styles in general include the details of the author, year of publication and page number. In-text citations are always recommended when referring to a sizeable chunk in the write-up using round or square brackets; "(...)" "[...]". The use of quotation marks is also suggested when a specific quote is included.

(ii) References: It is the list of the sources explicitly used in completing the write-up or the manuscript. It includes detailed depiction of the author, journal/proceeding/book, publisher, publication year, volume, issue no., and page no. (or date of download in case of multimedia sources).

(iii) Bibliography: This is a list of all the sources that are used in completion of the manuscript but may have not been directly used in the contents of the assignment/assignment. It is the list of all the references published/unpublished which have provided background information and provided better understanding of the topic/concept but are not explicitly mentioned in the manuscript. The Bibliography is obviously bigger than the reference list.

Diagrams, images, graphs equations and tables used in the article should also be acknowledged using appropriate and acceptable referencing style. It is a standard practice to acknowledge the source in the caption using parenthesis.

The descriptions used in the sources should not be replicated directly without any quotation marks and the corresponding reference's number, as it would lead to plagiarism. The quotation marks should not be used by extensively copying the data, as it would almost destroy the very purpose of writing a research paper. As previously explained, the quotation marks should be used with the appropriate reference number for descriptions that could not be explained any better than the source/reference itself.

For example, consider a statement; A research could be termed as the "processing of the information". The sources concerned with the research should be properly consulted. "Research is a dynamic cerebral activity, while reporting is a mechanical one". It should be ensured that a research paper is not biased and well cited. Citation is needed for both the above sentences.

In case of a scenario where an already existing concept is explained so well that it cannot be explained any better, it is allowed to use the same description. But, note that such a description should be enclosed within quotes and appropriate citation should be provided, else the research paper risks being plagiarized. Explaining an existing concept in the writer's own words is allowed, as it defines the understanding of the writer and his/her capability to summarize and phrase the "description". It would not lead to plagiarism "if the writer is putting the information in his/her own words as it demonstrates the understanding of the information" as far as the writer is concerned.

A report is sometimes good enough even though it merely includes data gathered from multiple sources, but a research paper leads to fruition only if some essence is extracted from the available data.

A research paper should be balanced as far as the academic arguments are concerned. As previously mentioned, a report is merely a collection of ideas, while a research is a thorough analysis of those ideas. A research paper should not merely be rhetoric, but it should be written in a clear and concise manner. It should be written in such a way that it is used extensively as a citation or as a source for other writers.

7.8.2 Paraphrasing

Some of the literature on "Plagiarism" categorizes paraphrasing as a type of plagiarism, whereas some of the literature describes paraphrasing as a technique to avoid plagiarism. This section presents both the cases considering appropriate examples.

Paraphrasing is expressing an already reported idea/contribution in your own choice of words or text. In such a scenario, where the author summarizes someone else's work by just changing the words, reframing them or uses synonyms retaining the essence of the already reported work is still categorized as intentional plagiarism. Consider the following example.

Original text:

> Results-based instruction has been the theme of caustic discussions in numerous states and educational systems. The tumult amazed a few instructors who, following quite a while of hearing calls for "comes about" from political and business pioneers, accepted that most guardians and nationals would bolster a move to more unmistakable results and methods for surveying them. Maybe the lion's share of individuals do concur with the standards of OBE – or would in the event that they comprehended them – however exceptionally vocal commentators have how brought up enough issues about how OBE may function by and by to make questions among educated individuals from the general population about its allure. Should instruction be result based? Some may contend that it as of now is, to some degree. Almost all training organizations have objectives that apparently direct their work. At the point when instructors design educational programs or educators design exercises for their classes, they typically begin by clearing up the reasons.
>
> (Source: *Curriculum Handbook: An Overview of Outcome Based Education*, by Ron Brandt)

Unacceptable Paraphrase or Plagiarism

> Outcome-based education has been the subject of discordant discussions in many states and school systems. This surprised some of the educators because they assumed that society would appreciate the students being assessed on definite outcomes and means of assessing them. Perhaps the majority of people were in favor of the principles of OBE, however some concerns were also registered among the members of the public. There is also ongoing debate on the issue that is the education outcome based with supports in favor and against. Nevertheless all the institutes are driven by goals, purposes and objectives.

In this example the author attempts to paraphrase or summarize the original text by using synonyms which ultimately is still considered as plagiarism, despite the fact that both the versions are not exact match of words. Acrimonious is replaced by discordant, debate by discussion and so on. Moreover some of the text in the original version is just rearranged to make it look different or unique. So the new edited version is not significantly unique and doesn't reflect the originality of the author's ideas. It is important to note that even if such a modified version is referenced, it is still a case of plagiarism because it is still remains a mere copy of the existing text or material.

Acceptable Paraphrase

> Outcome Based Education (OBE) is emerging as an inevitable dimension in the education system demanding a complete shift of teaching-learning paradigm. Ron Brandth, in the Curriculum Handbook has presented an overview of "Outcome Based Education (OBE)." The author has presented various dimensions of OBE highlighting the importance of definite outcomes achieved by the students and measureable means of assessing them from the prospective of the parents and citizens. The author also clarifies the role of goals set by the respective education institute which defines the expected outcomes and obviously the corresponding objectives of the courses.
>
> [25].

In this example, that demonstrates acceptable paraphrase, the sources referred by the author are precisely acknowledged and there is no attempt of taking credit of someone else's work and ideas. Moreover, most importantly as expected in a successful paraphrase, the author has integrated the entire context presented in the original source and expressed his/her understanding of the context in his/her own choice of words, retaining the attribution to original source. Such an attempt of paraphrasing is considered as an acceptable technique to avoid plagiarism. It is also imperative to note that even if it is considered as acceptable paraphrase, it is always recommended to provide footnotes and endnotes indicating the details of the broader ideas/sources or information used to write the article, assignment or manuscript. Due care taken by the author while preparing an assignment is directly proportional to the credibility/integrity and obviously inversely proportional to the probability of being accused of unacknowledged intentional/unintentional plagiarism. A genius writer also makes it a habit of using multiple sources, while paraphrasing. An approach of explaining a concept using various sources, providing further clarification, presenting the comparison of the views/findings/conclusions or perspectives reported by the various authors and finally going beyond just merely presenting the survey to add on your own opinion, identifying unreported/unaddressed issues and novel finding of the topic under the discussion, makes it a great article/assignment and makes you a genius author exhibiting highest degree of integrity.

With exponentially increasing need of paraphrasing as a tool to avoid plagiarism and with automation, touching almost every nook and corner of human life, it is absolutely; no wonder to discover internet flooding with the numerous free, online tools and software for automated paraphrasing and text rewriting. Some of them are listed here:

- https://paraphrasing-tool.com/
- http://www.paraphrasing-tool.org/

- https://www.prepostseo.com/free-online-paraphrasing-tool
- http://www.paraphrasing-tool.net/
- https://seomagnifier.com/online-paraphrasing-tool
- https://www.rewritertools.com/paraphrasing-tool
- http://www.goparaphrase.com/
- https://www.paraphrase-online.com/
- https://articlerewritertool.com/
- https://paraphrase.online/

7.8.3 Tips and Strategies to Avoid Plagiarism

Research writing is an art of giving something novel to academia. It should not merely be an amalgamation of facts and ideas from different available sources. The writer must have a thorough conversation with the sources from which they want to obtain information. They must delve deep into the subject they are dealing with. The research no longer remains one, if the facts from different sources are just merged into one unit. Unlike report, it renders information about an idea, rather than solely being a chain of few concepts. Educators must equip students to develop their own ideas or develop insights into existing ideas. Moreover, researchers should be readily encouraged to pursue researches. Every idea should be appropriately backed by an existing concept. It should not be written based on an opinion; rather some experiments should be conducted to support a claim. It is suggested to explore the topics of interest in great detail, put a genuine effort to conduct research in it and pen the inferences in the form of a research paper in such a way so as to not be subjected to plagiarism. Some of the tips are detailed to avoid infringement of plagiarism polices.

(i) Inculcate habit of writing down your own ideas, findings and conclusion and also taking systematic, comprehensive and effective notes of the material from all the sources. This will help the writer to:
- Develop clear understanding of citation and avoid plagiarism
- Evaluate the sources for their usefulness and accordingly acknowledge the relevant material
- Distinguish his/her thoughts from the reported or existing ones to highlight novelty
- Create an outline of the proposed paper/article/manuscript
- Formulate the approaches to fit the contents in the identified journal or newsletter
- At whatever point, you download material from the Web, make certain to make a different archive petition for that material. Web locales are more unstable than print sources. Material on numerous web destinations is refreshed once a day and a webpage that you find right off the bat in your examination might be passed when you compose your last draft. Hence, you ought to dependably record bibliographic data for every web sources as you utilize it

(ii) With an idea in hand, it is advised to follow a top-down approach. Breaking down the crux of the idea in hand for further details is suggested in a research paper. Writer should know the details to be avoided from the available information. The information which really matters for the research paper should be part of it. The details should not be reciprocated exactly; instead, the conclusion which the writer draws out from those details should come to light, else it turns to be similar to a report

(iii) Understand the difference between paraphrasing and summarizing:

A summary of an article is expected to include only key points and should highlight the important facts of the article. The exact words or some content may be replicated in the summary. It is akin to an abstract of a manuscript and therefore is always much shorter (lengthwise) than the original article. A paraphrase, on the other hand, may be about the same length as the original article. Rules already discussed need to be precisely followed during paraphrasing.

(iv) Understand difference between collaboration and collusion

> Almost everyone has difficulty identifying where collaboration stops and collusion begins.
> (Carroll & Appleton, 2001, p. 15)

Collaboration is encouraged in academia and other domains as well. However, if not carefully dealt, the threat of collusion outweighs the advantages of collaborative learning. For example, students are expected to work as a team, in a group for their final year undergraduate engineering project but are assessed individually for their technical knowledge and presentation skill during their project examination. It is imperative that the team mates are taught and guided to collaborate effectively, share ideas and jointly with synergy during the planning of the project and its execution. This collaboration and coordination should necessarily end at this stage and should not cross the limit while the students individually compile the report and present the results during the submission for individual assessment. Restricting the team work to legitimate discussion which demand contributions from all the team mates in the initial phase and later focusing on individual contributions will avoid chances of collusion. The scenario will be completely different with absolutely no chances of collusion if the same project group now participates in a competition where they are assessed as group and group as a whole gets the marks.

(v) Develop good academic, personal and literacy skills. This will inculcate habit of being disciplined, avoid laziness. The authors should also undertake courses to improve academic/technical/nontechnical writing and improving presentation skills

(vi) Study the expectations/rules and regulations under academic sections in your university or the corresponding domain where you aim to publish. The plagiarism policies are well defined and differ across the disciplines, various journals in a discipline, and also across various organizations.

(vii) Follow the terms of open source license that applies to publically available material. The copyright material must not be published (for example on a website) unless you have permission from the owner of the copyright.

(viii) Faculty are encouraged to create assignments that minimize the chances of plagiarism and emphasize on original thinking of the students. For example: An assignment on the course of Network Analysis may be framed as:

Design a series resonating circuit to have the cut-off frequency, Fc (in kHz), which is your *"Roll Call Number."* Assume the value of capacitor (in mF) which equals your *"Permanent Registration Number"* (PRN). Find the deviation in calculated cut-off frequency of the series resonating circuit if the bandwidth has to be changed to your CGPA score of first year.

7.8.4 Creative Commons License

Creative Commons (CC) is a worldwide charitable association that empowers sharing and reuse of innovativeness and learning through the arrangement of free legitimate devices. CC permit is one of a few open copyright licenses that empower the free conveyance of a generally copyrighted work. A CC permit is utilized when a creator needs to give individuals the privilege to share, utilize, and expand upon a work that they have made. CC gives a creator adaptability (for instance, they may permit just non-business employments of their own work) and ensures the general population who utilize or redistribute a creator's work from worries of copyright encroachment as long as they keep the conditions that are indicated in the permit by which the creator conveys the work.

There are a various types of CC licenses. The licenses contrast by a few blends that condition the terms of appropriation. They were at first discharged on December 16, 2002, by Creative Commons, a U.S. nonbenefit organization established in 2001. There have likewise been five variants of the suite of licenses, numbered 1.0 through 4.0. As of February 2018, the 4.0 permit suite is the most present.

CC0 empowers researchers, instructors, scientists, the contributors and proprietors of copyright- or database-secured substance to postpone those interests in their works and in this manner put them as totally as conceivable in people in the "public domain" so others may uninhibitedly expand upon, improve, and reuse the works for any reasons without limitation under copyright or database law.

(https://creativecommons.org/share-your-work/public-domain/cc0)

Creative Commons Attribution (CC BY): Provides opportunities to make abstracts, and other changed variants, adjustments or subsidiary works of or from an article (for example, an interpretation), to incorporate into an aggregate work (for example, a compilation), to content or data mine the article, notwithstanding for business purposes, as long as they credit the author(s), don't characterize the writer as embracing their adjustment of the article, and don't alter the article so as to harm the writer's respect or notoriety.

Creative Commons Attribution – Non Commercial – No Derivs (CC BY-NC-ND): for non commercial purposes, gives others an opportunity to disseminate and duplicate the article, and to incorporate into an aggregate work, (for example, a treasury), as long as they credit the author(s) and gave they don't adjust or alter the article.

7.9 Summary

Literary theft is a typical and major issue in the scholarly field and somewhere else. Copyright infringement in the scholarly world can happen in content or source code. It facilitates one"s errand at the cost of someone else. Much of the time, copyright infringement happens because of absence of appropriate affirmation of work done by others. This circumstance can be effortlessly maintained a strategic distance from by offering direction to understudies on the hugeness of incorporating appropriate references in their work. From this section we come to know different programming apparatuses and prescribed accepted procedures, the understudy, workforce, look into researchers or other individuals, can ready to present their papers or original copies un originality free. Be that as it may, even we come to know appropriate free and paid programming for testing counterfeiting, and consequently it is extremely advantageous for speaking to the nature of work. Subsequently, this empowers to stay away from literary theft in understudy and also proficient life.

Further Reading

1. Urkund website
2. https://timesofindia.indiatimes.com/life-style/fashion/designers/rahul-mishra-accuses-indian-ethnic-wear-brand-and-a-french-label-of-plagiarism/articleshow/64176354.cms
3. https://timesofindia.indiatimes.com/articleshow/65261081.cms?utm_campaign=andapp&utm_medium=referral&utm_source=whatsapp.com&utm_source=contentofinterest&utm_medium=text&utm_campaign=cppst
4. https://www.plagiarismtoday.com/2018/01/04/copyright-plagiarism-2018-2017/
5. Copyright and Fair Use Portal at Stanford University, http://fairuse.stanford.edu/, visited: 22 July 2006
6. J. Band, "The Google Library Project: Both Sides of the Story", *Plagiary: Cross-Disciplinary Studies in Plagiarism, Fabrication, and Falsification*, vol. 1, pp. 1–172006.
7. https://www.plagiarismtoday.com/2011/10/04/the-world%E2%80%99s-first-plagiarism-case/
8. COPE (https://publicationethics.org/)
9. MIT, (March 25, 2003), "Budget Projections, Student Discipline Report Presented to Faculty", http://web.mit.edu/newsoffice/2003/facmeet.html, visited: 22 July 2006
10. http://www.ascd.org/publications/curriculum_handbook/413/chapters/An_Overview_of_Outcome-Based_Education.aspx (example of outcome based education)
11. EVE Plagiarism Detection System website, http://www.canexus.com/eve/, visited: 22 July 2006.
12. The Stanford Daily, February 12, 2003 By A. Alemozafar, http://daily.stanford.edu/article/2003/2/12/onlineSoftwareBattlesPlagiarismAtStanford, visited: 22 July 2006
13. Berkeley University of California, Student Conduct, Sanctions, http://students.berkeley.edu/osl/sja.asp?id=1004, visited: 22 July 2006.
14. Glatt Plagiarism Services website, http://www.plagiarism.com/, visited: 22 July 2006.
15. www.centralia.edu/academics/writingcenter/
16. http://www.hygeiajournal.com/downloads/Editorial/1597787464plagiarism.pdf. *Hygeia: Journal for Drugs and Medicines*. ISSN 2229 3590 (online).
    K. R. Vinod, S. Sandhya, K. D. Sathish, A. Harani, D. Banji, O. J. F. Banji. "Plagiarism-History, Detection and Prevention"
18. J. D. Beasley, "The Impact of Technology on Plagiarism Prevention and Detection", Plagiarism: Prevention, Practice and Policies 2004 Conference.
19. https://www.tru.ca/arts/php/history/A_Handbook_for_TRU_History_Students/Plagiarism.html
20. Massachusetts Institute of Technology Policies and Procedures, http://web.mit.edu/policies/10.0.html, visited: 22 July 2006.

21. Stanford Honor Code, http://www.stanford.edu/dept/vpsa/judicialaffairs/guiding/pdf/honorcode.pdf, visited: 22 July 2006
22. https://en.wikipedia.org/wiki/Plagiarism
23. https://plagiarismsearch.com/blog/6-scandalous-plagiarism-stories-that-you-should-know.html
24. Plagiarism Tour at turnitin.com, http://www.turnitin.com/static/flash/tii.html, visited: 22 July 2006
25. https://timesofindia.indiatimes.com/topic/plagiarism

8
Intellectual Property Rights

Dipali Kasat, Rajeev Kumar, and Shailaja Patil

CONTENTS

8.1 Introduction 266
    8.1.1 Intellectual Property 266
    8.1.2 Difference between Industrial Property and Intellectual Property 267
8.2 Significance of Intellectual Property Rights 267
8.3 Forms of IPR...268
    8.3.1 Patent 268
    8.3.2 Copyright 279
    8.3.3 Trademarks 280
    8.3.4 Collective Marks 282
    8.3.5 Industrial Design...282
    8.3.6 Integrated Circuit..283
    8.3.7 Geographical Indications 283
8.4 Importance of IPR in Global Economy 284
    8.4.1 Valuation of IPR 285
8.5 Role of Intellectual Property in Technology Transfer 286
    8.5.1 IPR and Licensing 286
8.6 The Philosophy of Intellectual Property Rights................................287
8.7 IPR as an Instrument for Development 287
8.8 Some Examples of Patent 287
    8.8.1 Self-Healing Power Grid and Method Thereof (U.S. Patent No. 8504214) 287
    8.8.2 Infectious Disease Detection (U.S. Pat. App. No. 20130130227) 288
8.9 Summary 288
Further Reading..289

> Intellectual Property has the shelf life of a banana.
>
> —Bill Gates

Learning objectives of this chapter are to:

- Understand various forms of Intellectual Property
- Explain the strategic value of IP and how IPR is applicable in research

- Illustrate the role of Intellectual Property Rights in research and in transfer of technology
- Evaluate the importance of IPR in modern global economic scenario

This chapter will enable the researcher to:

- Understand the scope of IPR
- Predict and protect the research novelty in the form of IP
- Select the IPR tools that acquire legal rights of exclusively using creations of one's own for a certain period of time
- Judge research innovation from patenting aspect

8.1 Introduction

It is accepted universally today, that Intellectual property (IP) and intellectual property rights (IPRs) are important in the progress of nation development, and are considered as a barometer of research and development activities. The main aim of this chapter is to give researcher an experience of deep dive into the world of Intellectual Property. Following sections will help you to understand the strategic value of IP and how IPR can be applied to the research, innovations and the artistic creations by encouraging technology transfer. For a fast pace technological/industrial development an understanding of IPR is utmost essential.

8.1.1 Intellectual Property

There are various forms of invention, creativity, research, innovations, artistic creations that are covered by Intellectual property. In daily life, we come across various forms of IP like, literary work (e.g., Book), electronic devices (e.g., Mobile), songs, movies, pharmaceutical products etc. The legal rights covering IP are called as Intellectual property rights. For example, an original literary work and an invention are intellectual property and when protected by the IPR, they will be termed as "copyright" and "patent," respectively.

There are two types of properties:

1. **Physical Property:** It is the tangible property such as Land, Ornaments, House, Money, Building, and so on.
2. **Intellectual Property:** It is the intangible property such as Patent, Copyright, Trademark, Design, Geographical indications, and so on.

Like physical property, intellectual property can also be protected by registering it under an appropriate legal regime. Intellectual property laws differ in each jurisdiction. The implementation of IP rights obtained distinctly in each territory of interest, is directed by the rule of the country in force, namely the Law of Intellectual Property Rights.

IPR is a tool for the owner of IP providing exclusive rights in the form of Copyright, Patent, Trademark, Integrated Circuit, Geographical Indication, or Industrial Design. The owner of IP acquires legal rights of his/her creations for a term period, known as Term of IPR. Table 8.1 provides a summary of the types of intellectual properties along with the duration, called as Term and the subject matter.

8.1.2 Difference between Industrial Property and Intellectual Property

The word Property authorizes those articles that are normally acknowledged to be an asset of a single entity or a group. For the ease of understanding, the IP can be categorized as, Industrial Property and Copyright. Industrial Property refers to the industrialized intellectual yields, such as patents, designs, trademarks, trade secrets, and so on. In addition to Industrial property, the Intellectual Property includes other rights such as copyright for literary works, artistic works, software etc. However, today, it is common practice to use the word Intellectual Property as a catch-all term that comprises of all properties resulting from the practice of human brainpower.

The colloquial definition of Intellectual Property can be stated as it is that intangible property that is coming from the exercise of human intellect, such as inventions, ideas, music, design, article, poems, and so on. In short, Intellectual Property is about creative ideas.

There is a wide range of creations covered in Intellectual Property Rights, such as a copyright on an article or book, a trademark, which is a logo that distinguishes its logo design signifying a firm and its services, a patent on the invention or innovative design/process, various trade secrets, geographical indication such as Darjeeling Tea, trade-related aspects such as trade secrets, and so on.

8.2 Significance of Intellectual Property Rights

In the modern business model, concepts and understanding are important from the business point of view. A major worth of pharmaceutical and cutting-edge technology products lies in the extent of innovation, research, design and testing, invention involved.

TABLE 8.1

Forms of IP with Subject Matter and Term

Form of IP	Subject Matter Protected	Term of IPR
Patents	Inventions	20 years (from the date of priority)
Industrial Designs	The aesthetic or ornamental aspect of useful objects	15 years
Trademark	Indications to distinguish the source/origin of a product or service	10 years from the date of registration
Copyright	Literary, artistic, dramatic, and musical works	60 years
Geographical Indications	Indications of geographical origin where special features are attributable corresponding to the territory	10 years
Integrated Circuits	Layout designs	10 Years

For business growth of any enterprise today, it is important to protect the novelty in the form of IP because it is a key element required to maintain a competitive point in the marketplace. IP is an integral part of the commercial process. The supervision and protection of IP can mean the difference between victory and disappointment in business today. The Movies, music rendition recordings, theater renditions, books, computer applications, and online facilities are basically traded due to the originality of content. Many commodities that used to be imported or exported, now contain innovation and design in their goods, like brand-named products/clothing. Therefore, inventors are given the Intellectual property rights to inhibit others from using their foundations, inventions, or designs. Thus, IPRs are the key handlers of economic performance in Research and Development–based business models by preventing others from infringing the original creation. It is to be noticed that rationalization of the implementation of the rights ensued is restricted to the territory/country which permits such rights and cannot be extended beyond the province of the that country.

8.3 Forms of IPR

World Trade Organization (WTO) recognizes following forms of IPR, under Trade-Related Aspects of Intellectual Property Rights, known as TRIPS agreement [1]:

- Patents
- Copyrights
- Geographical indications
- Integrated circuits
- Trademarks
- Trade secret
- Designs

In addition to all this, there are other forms such as plant variety protection, utility model, and so on. However, the important and common forms of IPR include Patents, Copyright, Trademark and Design. The Patents, Copyrights, and Designs all fall under the category of IPRs that encourage inventive and artistic works. Trademarks and Geographical indications formulate the informative category of IPRs. Figure 8.1 summarizes the forms of intellectual property associated with the IPR.

8.3.1 Patent

The word "patent" is originated from a Latin word "patere," which means "to lay open" (*i.e.* to make accessible for unrestricted assessment). Patent is a form of IPR attained by an inventor for his/her technological invention. An invention can be termed as a novel product or procedure involving an innovative step and capable of industrial applications. To be patentable, an invention must be Original, Nonobvious, and Industrially applicable.

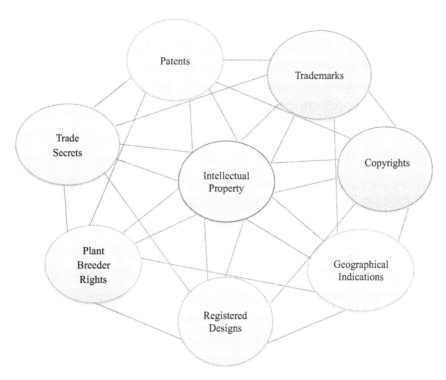

FIGURE 8.1
Forms of Intellectual Property Rights

- *Original:*
  - not previously known or used by anyone in the public domain
  - not available in any document of any publication in the electronic or print media (no publication even of the inventor as author prior to filing a patent application)
- not a part of the state of the art *Nonobvious:*
  - should include the element of technological advancement and economic significance
  - have a known use or produce a concrete and tangible result not clear to anyone having average skills in that area of technology allied to that invention
- *Industrially applicable:*
  - Capable of being manufactured and used in the industry. In broader sense, the term "Industry" is applicable to all other fields also like medical, agriculture, public services etc. It also means, invention should be infeasible for non-patentable subject matters such as perpetual law of motion, methods of medical treatment etc

The inventor can check these three features, viz. originality, nonobviousness, and industrial applicability of his/her invention by doing a thorough prior art search. To do a prior art search the inventor must

- Derive the outline of the invention and identify the keywords with all the possible synonyms
- Think about where most relevant information would be. A lot depends on the nature of the invention
- Search journal articles and research papers
- Go through books related to the discipline
- Check on Internet search engines (e.g., Google)
- Search the prior patents in commercial databases and national patent office databases
- Patent Databases
  - The Indian government allows the public to freely search on their Web site, ipindia.nic.in. [2]. Additionally, other national offices allow you to search their databases
  - European databases on patent [3]
  - Japanese database on patent (in English): [4]
  - The United States Patent and Trademark Office (USPTO) website: [5]
  - World Intellectual Property Organization (WIPO) website: [6]

After prior art search with reference to above links, the inventor should decide the category of his/her invention. An invention to be patented falls in one of the following categories

- Utility: process (method), machine, manufacture, or composition
- Design: ornamental design for an object, Industrial design of the product
- Plant: new variety or any new and useful technological improvements

(i) Utility Patent: The former and most regular patent is known as a utility patent. A utility patent ensures the way an item is utilized, merchandized and it is widely recognized. The other type of patent one can acquire is known as a design patent also called as outline patent, which secures the way an object looks. The last type of patent one can get is known as a plant patent, as the name suggests, it ensures disclosure of novel assortments of plants.

After the prior art search and decision of category, the inventor can then proceed to file a patent application. Patent filing specification can be either provisional specification or complete specification. If the invention is still at a conceptual stage, the inventor can secure his/her idea by filing the provisional patent or else the inventor secures the invention with complete specification. But legal rights are granted for complete specified patent only; provisional specification is preferred just to fix the patent's date of priority. The Provisional patent application is used to secure the idea and the inventor gets a period of twelve months to submit the complete details of the invention.

(ii) Who can apply for a PATENT?
Patent must be connected with the names of the "innovators" who have contributed in the development of patent. In the event that any individual has made some commitment about the origination of one component in any claim recorded in the patent application claims, at that point that individual must be recorded as a designer in the patent application. It doesn't make a difference whether you have contributed 99.9% toward the origination of your creation. In the event that someone else has contributed the

remaining .1% toward the creation, at that point you should apply for the innovation with that individual if that .1% is a piece of any entitlement in his/her application. There can be combined inventorship if two or more participants work toward the invention.

Wherever two people are absolutely ignorant of each other's work respectively, regardless of whether they are utilized by a similar organization, there is no combined inventorship.

(iii) Benefits of applying for a patent

A patent is a legal protection that gives an inventor monopoly and the right to exclude others from doing certain endeavor in the country of issuance. The international norm is to provide a term of at least 20 years from the date of submission of a patent application. Patent rights prevent others from commercially using, making, importing or selling, or distributing it. Similar to any other property right, a patent may be licensed, transferred, sold, or simply abandoned. However, on the grant of patent, a patentee enjoys all the rights conferred by the law. The patentee can setup a venture, launch the product in market and gain profit till the term ends. There is a hard fact about patent, beautifully quoted by Mark Twain, "The man with a new idea is a crank until the idea succeeds." The protection ends once a patent terminates, and such invention enters into the public domain, wherein the proprietor holds no exclusive rights to the invention. Such a patent can be used by any other interested person or entity for manufacturing and distributing to be used in favor of public at lower price. A very good example of such case can be seen in generic drugs. After the completion of term of any pharmaceutical product, such drugs are manufactured and sold as generic drugs at a price that is affordable to general public.

(iv) Common consideration in patent filing

An idea with potential commercial application needs hard efforts of a researcher. But it takes lot of efforts, labor, time, and resources to get that idea in one's name in the form of "Patent". Getting an invention protected by a patent needs continuous patience and yet hard efforts. Apart from investing in time, money and resources in the patent filing process, one must spare some time understanding the pitfalls that one can fall prey to, inadvertently, in the process of getting a patent granted. A little more effort to develop an understanding of the "Invention Industry" can put you on the safe track, right from the beginning of patent protection idea.

Following are the few prerequisites for patent filing:

Getting an idea and its preliminary success cannot assure it as a patent. There are a lot more check points which an idea has to successfully cover in order to assure its claim for protection as patent.

- **Novelty**

  The idea must be novel in every sense. The science or the skill might be well versed with the world but the art or its use must be new. This is the very first step in patent filing process which needs a lot of time and capital investment. A careful and intelligent approach can help to cross the first barrier on whose basis preliminary screening of a patent application is done. It is very encouraging to look oneself for the similar inventions or ideas in literature through internet. Such self-searching rules out the major apprehensions about the novelty of the work. But relying completely on the Internet self-search can put the invention in trouble as there can be chances that some of the filed applications similar to the intended invention might just not come up as search result. There is a problem with most of the inventors who are so enthusiastic about their invention or idea that they want to do it all by themselves. This might be for saving capital resources but this small act of money saving can make the application prone to cancellation on a novelty basis.

- **Utility**

This is very important for an invention to find utility in society. There is no use of an invention or idea that cannot serve in public. Finding the utility is sometimes overlooked by inventors. One must be clear about the section of society whom the invention is going to benefit. This not only widens the horizon of initial idea and concept but also adds to the clarity of the scope of your invention which is to be mentioned in Provisional Patent application. A proper market survey also helps to know about what edge your invention has over the pre-existing products in the market. A true picture of marketability of your invention or product gives you a clearer insight to make the deals with product licensing firms in future who will take up your product to the market and save you from making wrong choices or losing profitable deals.

Once the invention or idea passes all of these mentioned prerequisites, it is considered to be ready to put into the patent filing procedure. The degree and amount of hard work that brought the invention in reality is what is again sought, to get the invention protected through patent. One is expected to be technically aware of all the legal terms and conditions of "Invention Industry" that are to be followed throughout the journey of "Patent Granted" status. The two main terms are:

- Date of conception of idea/invention:

This date has got a great significance as it marks the time since you claim the invention. This is not just the date when the inventor started working on the idea but the date when inventor owned the complete functional idea of the invention.

- Selling the invention or first public use:

Once the invention is sold or is used publicly, the significance of date of conception in patent filing application is carried away by the *date of first public use* of invention after it is sold.

According to USPTO, in 2011, it is not just enough to be *"first to invent"* as per the date of conception of invention. They introduced *"first to file"* rule and made it important to complete the non-provisional and provisional filing process to claim for any invention. The section below describes some important facts to be considered by the inventor while filing a patent application.

(v) **Do's and Don'ts of Patent Filing Procedure:**

- Protecting intellectual property:

Having known the novelty and market value of the invention, it is apparent for any noble inventor to go and talk about it among peers and society. Though done for bragging or to grab some more good ideas for improvisation, this act is generally the most avoidable in case of patent filing. The invention that is intended to be filed for patent must be kept secret for as long as possible. By keeping secret, it is meant to not to disclose the invention in any form that might lead to its copy or even improvisation. It is wiser to stay silent till you get it. It is very important to protect your idea till patent is filed.

- Time limit for patent filing (or the 12-month window)

The USPTO or any patent granting organization gives a particular time span (12 months for USPTO) starting from the date of conception of invention or date of first public use, after which the patent file is not considered for granting the patent. Missing the last date to file the patent can, in the first place, end the possibility of getting a patent granted.

- Drafting claims in filing patent

The idea of invention must be clear and informative enough that a novice too can understand. It must reflect a perfect understanding of the inventor about the functional details of the invention. The description of invention at every step of filing the patent (be it primary non-provisional application or the final provisional patent application) should be same and equally supporting to the expected product of the invention. Any mismatch between the facts or description presented about the invention in applications or the final product may lead to the cancellation of the application. This means starting all over again.

- Assurance

The popular phrase "haste makes waste" makes perfect sense with inventors when it comes to a new invention. Any sort of act to make money or recognition can put the fate of patent on stake. Rest assured by filing the application for patent and selling the invention even before the patent is granted, can be a disaster if, there occurs any discrepancies with the Provisional Patent application resulting in cancellation of file. This puts you again on step 1 of patent filing which now considers the date of first public use of invention. In case of refiling the patent, there are increased chances of crossing the required time limit for filing the patent and losing the opportunity to ever get the invention protected with your name.

- Clarify inventorship and ownership

When there are two or more persons involved in an invention, it becomes very important to clarify the ownership rights of the patent to protect the commercial value of the patented invention, in case of any dispute between the rightful inventors of the invention. Any failure or delay in reaching to an agreement can affect the market value of the patented invention.

- Diligence in record keeping

Getting an invention protected by patent in your name is a long and time-consuming process that requires an additional diligence in keeping the record of each and every detail. Any negligence in keeping records and verifiable proofs of date of conception of idea can endanger the viability of otherwise granted patent in case of any litigation against your application or claim.

A careful understanding of these points can be instrumental in strengthening the patent filing process and saving of time, energy, and resources of an inventor who intends to protect the intellectual property through patent.

(vi) Patent Application filing in India

The Patent Act, 1970 governs patents in India, which was modified in the years 1999 and 2016, respectively. A patent is a lawful control conferred to the owner of an invention for a specific duration. With this right, a patentee can stop others from selling, importing, or manufacturing the patented invention.

The time-span of examination and granting of patent varies depending on number of applications received at Patent office. Once a patent is granted, a patentee can enjoy its benefit for 20 years from the date of filing an application. The following Figure 8.2 shows the patent application process in India.

(vii) Application of Patent:

In harmony with the requirements of Patent Act, the inventor, his/her assignee or legal agent of deceased person, can apply for patents. Application can be made at the Indian Patent Office or its branches according to the territorial jurisdiction.

Patents can be filed online in respective jurisdiction: Mumbai, Kolkata, Delhi, and Chennai.

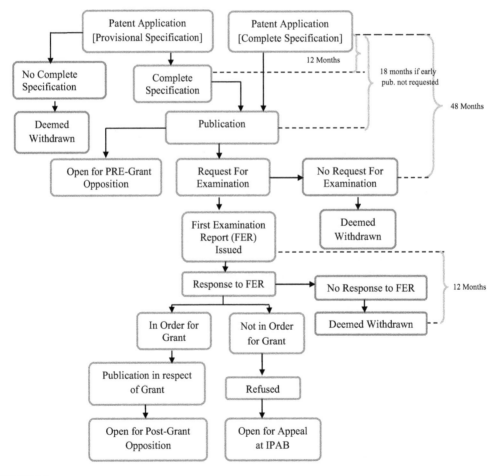

FIGURE 8.2
The Patent Filing Process in India

(viii) Regional Patent Office's Jurisdiction

The forms and documents that are mandatory for filing a patent application in India are as under:

- Form No.1- Application
- Form No.2: Provisional or complete specification
- (The complete specification must be filed within 12 months from the date of Provisional Application filing)
- Form No.3 – Statement of undertaking
- Form No. 5: Declaration of inventorship
- Form No. 26: Power of attorney (if filed through Patent Agent)

Online filing option is also available to inventors by using E-Filing Portal of the Patent Office. (https://ipindiaonline.gov.in/epatentfiling/goforlogin/dologin) Indian Patent Office's and their territorial Jurisdiction are shown in Table 8.1.

(ix) Types of Patent Application

Following are the types for filing national and international applications

A. National Application

- *Provisional application:*

Provisional application is basically for securing the inventors of idea and methodology. The provisional application contains description of the creation beginning with a title adequately indicating the invention related subject-matter, and drawings, wherever required. It should be noted that a Provisional Specification can't be submitted in case of Convention or a PCT National Phase Application. Here, filing a Complete Specification is compulsory. The thorough description should be submitted within 12 months from the priority date of the provisional application. If this requirement is not fulfilled, the claim shall be believed to be abandoned as per Section 9 (1) of the Indian Patents Act, 1970.

- **The components of a Provisional Patent filling**

There are two broad divisions of provisional patent specifications: Title and Description.

Title: Provisional patent specification begins with the title of the invention. The word limit for title is fifteen.

TABLE 8.1

Indian Patent Office's Jurisdiction

Patent Office Branch	Territorial Jurisdiction
Mumbai	The States of Maharashtra, Goa, Gujarat, Madhya Pradesh, Chhattisgarh, The Union Territories of Daman and Diu, Dadra and Nagar Haveli
Chennai	The States of Andhra Pradesh, Telangana, Kerala, Karnataka, Tamil Nadu, The Union Territories of Pondicherry and Lakshadweep
New Delhi	The States of Delhi, Punjab, Haryana, Rajasthan, Himachal Pradesh, Uttar Pradesh, Jammu-Kashmir, Uttarakhand, and the Union Territory of Chandigarh.
Kolkata (Head Office)	The rest of India.

Description: The description should begin with the preamble and contain the sections of field and object of invention. The technical field of invention should be mentioned in the field. It should be described aptly so that the nature of the invention and classification of its technology can be easily identified. The section of object is describing the necessity of the invention. It is essential to mention, what advantages the invention would bring as solutions to the existing issue.

- *Complete Specification Application:*

A complete specification is a document, which describes all details of an invention in a sufficiently clear and concise manner. The complete specification is different from that of the provisional specification. The complete specifications conclude with a set of claims. The important aspect of specification drafting is that, each claim in the complete specification may define an invention. Hence, a priority date is associated with each claim. The necessary forms required are already mentioned in above section. In addition to these forms, the applicant can request for early publication in Form No. 9, wherein the patent will be published within a month. If such request is not received, then the patent is published after 18 months from the date of filing. Note that, inventor may not be necessarily applicant. In such cases the applicant owns the patent.

A request for examination should be made in Form 18, within 48 months from the priority date. In case of nonreceipt of such application, patent application will not be examined.

- **Components of complete specifications:**

A complete specification should be submitted in Form No. 2. As mentioned earlier complete specification must contain the claims. The components of complete specification are as follows:

  - Title
  - Preamble of the invention
  - Technical field
  - Background
  - Objects
  - Brief description of the drawings
  - Detailed description
  - Claims (for complete specification)
  - Abstract

B. International Patent Application

1. International Patent application can be filed in two ways, Paris Convention
2. Patent Cooperation Treaty (PCT).

- **Paris convention**

The Paris Convention is first international treaty, signed at Paris in 1883. In Paris Convention inventors are allowed to file a patent in their own country first, and they get a time period of 12 months (after its initial filing date) to file elsewhere in other countries. This gives innovators extra time to perfect their product (with respect to need of market).

Within this time span one can also look for funding options. The Paris Convention allows inventors to use priority date of his/her own country while filing the patent in other country (within twelve months only). It saved the inventors time and money from having to file patents simultaneously in all the countries that he or she wishes to file.

- **Patent Cooperation Treaty:**

The PCT is an International patent law treaty that provides security for his/her invention in many countries. PCT is administered by World Intellectual Property Organization (WIPO). As of now, there are 152 countries that are signatories of PCT [7, 8].

Through PCT, it is possible to get patent protection for an invention in all signatory countries simultaneously, by filing an international patent application. Residents of India can file an international application with the Indian Patent Office at Mumbai, Delhi, Chennai, and Kolkata as the Receiving Office or the International Bureau of WIPO as the Receiving Office.

- **The PCT procedure**

The PCT procedure consists of two main phases. The international application is filed in first phase and the international application enters into PCT designated countries, where the application is evaluated under the patent laws. Thus, there are two phases: "international phase" and a "national phase."

 o International Phase: It is the phase before the entry of international patent application into designated countries. During the international phase, the international patent application is published, searched and examined

 o National Phase: The National Phase follows the International Phase and involves in finalizing the application in the Patent Office of specific countries. It is essential for an applicant to submit a nationwide phase application in correspondingly selected nation, where patent protection is requested for, within 30/31 months from the earliest filing date of the initial application. The granting/nongranting of patents is decided by the national or regional patent Offices in what is called the "national phase"

Figure 8.3 shows the international filing system and its timeline.

Procedure for PCT requires application consisting of a request containing description, claims, abstract, drawings of the invention. The language can be English or Hindi. The international application needs to be filed in triplicate. The PCT procedure includes:

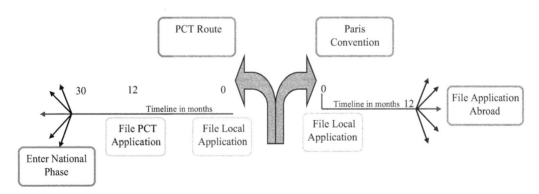

FIGURE 8.3
International Filing System

- Filing: Application can be filed in regional patent office
- International Search: An International Searching Authority (ISA) verifies the published patent application and prior art that may have an influence on whether your invention is patentable, and gives a written opinion on your invention's potential patentability
- International Publication: On expiration of 18 months from the filing date, the patent application is published
- National Phase: Usually at 30 months from the filing date of application, from which priority is claimed, the grant of patents can be pursued directly at national patent Offices of the countries where the applicant wants to obtain the grant

(x) Oppositions:

There are two types of opposition that an application might face Pre-Grant and Post-Grant Opposition.

- **Pre-Grant Opposition**

The Section 25(1) of Indian Patent Act 2005 (amendment) discusses about filing a pre-grant opposition against a patent application. Under this provision, the patent application can be challenged by any person, or third party. An application can be submitted to the controller of patents in writing regarding an opposition to the published patent, before the patent is granted.

Pre-grant opposition are like safety net that captures objectionable patent applications before grant of patent. The opposition can be filed on the following grounds, as mentioned in Patent Act, Amendment Act, 2005 in Section 25(1)[(a) to (k)]

- If the invention is obtained wrongfully
- Lack of inventive step or obviousness
- Nonpatentable subject matter
- Prior public knowledge or public use in India
- Prior publication anticipation
- Prior claiming or anticipation in India
- Insufficiency of description of the invention

- **Post-Grant Oppositions:**

As the name suggests, the opposition can be filed even after the patent is granted. A post grant opposition can be filed within one year from the date of grant. A patentee has to justify that his/her invention is genuine and doesn't fall under any of the accusations made on the grounds, as mentioned below. A patent can be opposed on various grounds as mentioned in the Section 25(2) of the Indian Patent Act, 1970, as enlisted:

- If a patentee wrongfully obtains the invention from an opponent
- The patent claim/s are published in India or elsewhere before the date of priority
- The invention claimed by patentee was known publicly in India or elsewhere
- The invention is obvious or lacks an inventive step

- The invention includes nonpatentable invention
- The complete specification does not reveal the methodology used for invention or wrong information of the source; or biological material's geographical origin
- Failure to disclose information to the controller, regarding any applications that might be filed in any foreign country for the same invention
- Not filing the convention application within the 12 months from the date of priority
- Invention anticipated with traditional knowledge from India or elsewhere

The grounds of both, the pregrant and postgrant opposition proceedings are similar. This ensures a two-tier method of opposition to check frivolous patent claims and wrongful monopoly of the patentee.

(xi) Patent Renewal/Maintenance Fee
As per the Indian Patent act, the patent remains in force for 20 years from the date of filing with the payment of renewal fee or annuity fee. To maintain the patent rights and enjoy the ownership, patentee has to pay the fee each year from the third year to 20 years (Rule 80[1A,Patents act]).

The applicable fee varies depending upon natural person, start-up, and small entity; and also whether e-filing or physical filing is done. The renewal needs to be done every year before the end of grant date. For example, the restoration cost/fee may be paid before completing the fifth year from the date of grant, for the sixth year (in advance). The applicable patent renewal cost to pay can be found on patent official websites [9]. In case of inability to pay, an extension period of 6 months is available. Such ceased patent can be reinstated, by paying the renewal fee within 18 months from the date on which the patent ceased to have effect.

8.3.2 Copyright

Copyright is the right given by law that protects creativity and expression of ideas of authors, artists, etc. for their work. Copyright includes expression of ideas in an original way, in the form of poetry, book, drama scripts, music and other forms of artistic works like choreography, sculptures, paintings, drawings, photographs, and so on. Copyright protects the expression of idea but not the idea. The types of copyright works can be summarized as:

- Original literary work includes any form of published or unpublished, nondramatic textual works, databases, computer programs, and so on
- Dramatic work such as plays, dramas, and screenplays
- Artistic work such as painting, architecture work, craftsmanship, a sculpture, a drawing including a chart or plan, diagram, map, photograph, and so on
- Musical works
- Sound recordings, cinematography, or broadcasts

Following rights are exclusive to the copyright holder and consent is needed to:

- Create copies of the work
- Circulation of copies of the work

- Present the work publicly (such as for plays, film, or music)
- Exhibit the work publicly (such as for artwork, or any solid material used on the Internet or television)
- Derivative works including making alterations of the original work
- Copyright owners can give diverse users unlike rights from this list. The same image can also be used on a glossy magazine cover, in a commercial or included in a documentary film, for example

To file a copyright the author can follow the stepwise approach:

- Categorize work in appropriate category
- Determine the ownership
- Identify the application forms
- Fill the registration form online/offline as desired
- Pay the fee and submit the required documents

Copyright infringement is an act wherein the other party acts as having the exclusive rights of an author without the consent or authorization of the author. One cannot assume that any work on the internet is in community purview. Altering an image is an exclusive right of the copyright owner and requires permission almost all the time.

Any use of copyrighted material will not be considered as infringement when these are used for reproduction in news reporting, teaching, research, criticism, national cause, religious programs, and so on.

8.3.3 Trademarks

A trade mark is an identity that distinguishes goods and services of one from others. There are many types of trademark that receive protection under Lanham act. A common trademark contains word, phrase, name, logo, or symbol to identify source of products or goods. In addition, trademarks have other forms such as certification mark, collective mark and service mark, trade dress, and so on.

- **Trademark:** A mark that includes any combination of words, letters and numerals, logo, device, to form a label, signature, colored marks, and three-dimensional signs. Trademarks also includes audible signs (sound mark), olfactory mark (smell mark). It is an identity that enables a customer to identify the foundation or source of a product or facility

The unregistered trademarks are depicted by the symbol "TM". The registered trademark is denoted by the "®" symbol. A registered trademark is supposed to guarantee the quality of goods bearing the mark. It can be used as for marketing and building a brand.

The attributes that a trademark must possess are: it must be an invented work, and it must be distinctive. Also, the mark must not be deceptively similar to an existing mark, name, or surname of person, word that reveals the nature of product, or geographical name. Following the examples of trademarks depicted in Figure 8.4 uses either Word or Logo Mark ()

FIGURE 8.4
Some Examples of Trademarks

- **Service Mark**: A service mark is used to distinguish the services provided by any enterprise. The service mark is identified by symbol SM as opposed to TM. The line of difference between the SM and TM is very thin. The SM is used for services, whereas TM is used by enterprise to identify and distinguish products. The SM can be a system supplying public needs such as electricity, water, transport, communication, maintenance or repair work, or any such service. There are many enterprises those manufacture products as well as provide services such as Amazon. A very prominent example of service mark for restaurant services is McDonald's. Similarly, Dmart and Walmart are registered service marks for retail store services (depicted in Figure 8.5)

- **Certification Mark:**

The certification mark does certify the origin or nature of the goods and services. The certification mark is defined in The Indian Trade Marks Act, Section 2(1) (e) as:

> a mark capable of distinguishing the goods or services in connection with which it is used in the course of trade which are certified by the proprietor of the mark in respect of origin, material, mode of manufacture of goods or performance of services, quality, accuracy or other characteristics from goods or services not so certified [10].

When any product successfully passes a test, it is said to meet certain standard imposed by the certification authority.

The CM shows the level of standard of a company, that is, it shows that goods and services of the trader are certified as meeting certain standards laid by certification agency. With this, the consumer is assured that the manufacturer has ensured the standard of production. The certification mark identifies the material, origin, and the

FIGURE 8.5
Some Examples of Trademarks

quality of the goods and services that distinguishes a specific brand/company from other competitors. In short, certification marks define "standard" of goods and services.

Example: ISI mark used for industrial product, Woolmark is certification for fabrics, Agmark for all agricultural products, and so on. Figure 8.6 depicts examples of certification marks.

8.3.4 Collective Marks

As the name suggests, the collective mark (CM) is associated to a group and not any product or service. The Indian Trademarks act 1999 (Section 61) defines collective mark in Indian context.

Such trademarks are owned by some association, institute or organization. These trademarks are primarily owned by an organization, institutes or any association that is related to several members.

The CM is used by a group, who wants to be identified by quality and accuracy. This also includes geographical origin or other specific features set by the organization in which they are working or are related with. Many traders use the collective mark, when they are associated with that organization. It indicates the membership of the association, union, or organization.

Examples of collective trademarks are depicted in Figure 8.7. "CA," which is used by the Institute of Chartered Accountants; 2. "CPA," which is used by Society of Certified Public Accountants

8.3.5 Industrial Design

To protect the aesthetic aspect of any design, the Patent Office confers an industrial design right. Industrialization is the main cause for growth of protection of designs. The Indian Patent and Design act was passed in 1911 and amended in 2000. The act defines the design as "features of shape, pattern or ornament, lines or color applied to any object" [11].

The main principle for acquiring protection rights for a design is uniqueness and creativity of the design, for example, design varieties applicable to tooth brushes, furniture, electronic devices, jewelry, and so on. The creator should protect an industrial design generated from his/her creative thinking because keeping an industrial design is an incentive for their creativity and adds commercial value to a product. When the marketplace is loaded with a large variety of products with exactly same functionality,

FIGURE 8.6
Some Examples of Certification Marks

FIGURE 8.7
Some Examples of Collective Trademarks

industrial design is the key of choice in factors that magnetize consumers to a product or leads them to select one product over another.

However, the industrial design right protects exclusively the nonfunctional aspects of an industrial product but not technical features of the object. Once the design is registered, the design remains in force for 15 years.

8.3.6 Integrated Circuit

The protection of an integrated circuit refers to the protection of layout design of the circuit.

Every electronic device contains integrated circuit. With advent of technology, the semiconductor industry is developing very fast, and huge number of components is being accommodated on a single chip. Here, layout design of a semiconductor integrated circuit plays a very important role. Hence, layout design of integrated circuit is included in IPR. India is a signatory of the World Trade Organization, which follows the "Semiconductor Integrated Circuits Layout-Design Act" passed in the year 2000, under TRIPS agreement [12].

This act defines what is meant by the layout design and Semiconductor Integrated Circuit. The layout design is defined as, "a layout of transistors and other circuitry elements and includes lead wires connecting such elements and expressed in any manner in a semiconductor integrated circuit" [13].

The semiconductor integrated circuit is defined as "a product having transistors and other circuitry elements which are inseparably formed on a semiconductor material or an insulating material or inside the semiconductor material and designed to perform an electronic circuitry functions" [13].

An application for layout design protection can be made by the creator of the layout design or any legal representative. The application should be filed at any regional office corresponding to territory. A Term of Protection is 10 years for layout designs.

8.3.7 Geographical Indications

A geographical indication (GI) is a mark used on products that have a specific geographical origin such as country, region, city, and town; and possess abilities or a repute that are due to that origin. GIs have been given status of intellectual property because the product gets more commercial value by associating it to a particular place.

The Geographical Indication is defined in TRIPS Agreement, Article 22 as "indications which identify a good as originating in the territory of a Member, or a region or locality in that territory, where a given quality, reputation or other characteristic of the good is essentially attributable to its geographical origin" [14].

In India the protection of GI was enacted in 1999, as it was the member state of the TRIPS Agreement. Advantages of Registration and Protection Act states that Consumers are protected from deception by inhibiting misuse of GI by any unauthorized person and interest of producers is maintained. Moreover, the export market will have goods of Indian GI.

Geographical indications are goods such as natural, agricultural, or manufactured goods, where its features are attributable to its geographical origin, such as climate, soil, and a traditional method of production. Some examples of GI are Alphonso mango, Nagpur Oranges, Surat Zari craft, Chanderi Saree, Kanjeevaram Silk, Feni and Scotch whisky, Darjeeling Tea, Champagne, and so on.

It is important to note that, a single product such as a cell phone can have multiple IPRs associated with it. The new features invention can be patented. The device software can be copyrighted. Fashionable or stylish design of body of cell phone can be protected under Industrial Design. The logo on the device can have a trademark of a manufacturer.

8.4 Importance of IPR in Global Economy

The IPR have played a significant role in enhancing the scientific advancements worldwide and these rights have helped industries of all types (small, medium, and large) to grow up. There is direct role of IPR in facilitating economic growth, however, this topic has to be looked at by different perspectives for developed and developing nations, for national and state economies, and so on. There may be cultural differences or socioeconomic variations prevailing in same state as well. IPR not only induces economic growth but also creates enticement framework that inspires research, innovation, and development. More importantly, in today's world where globalization has taken over, IPR also helps to give remedies to various global challenges, so that increasing demands of food, pharmacy products, alternative energy sources, climatic challenges, and so on can be easily coped up.

The United States has set an example of IPR related economic growth; therefore, this topic has grabbed the attention of rest of the world and many studies are being conducted worldwide, on IPR related laws, policies, and theories to study the economic growth because of IP. Likewise, enforcement and protection of IPR is also important for international trades. Towards approaching this task, where economic growth was related to IP, the research was carried out by Rapp and Rozek in year 1990 [15], which is known as RR approach. The RR approach dealt with compulsory licensing, patents for pharmaceutical products, and so on; however, this approach did not take into account enforcement laws and effectiveness [16]. However, in 1977 Ginarte and Park used the RR approach and studied more exclusively five more aspects to compare patent laws of different countries including duration of protection, level of coverage, and so on. Maskus and Penubarti gave an empirical mathematical relation, which deals with the IPR and GDP.

Furthermore, it is very important to state that along with other parameters, foreign direct investment (FDI) also plays a major role in the economic development of country. It has been observed that an increase in the rate of innovation can be directly correlated to the increase in the rate of economic growth. However, IP solely is not responsible for economic growth; other important facts such as country's research base, trading policy, stable macroeconomic conditions, etc are also required for sustained growth in economy through inventions.

The role of IPR in economic growth is further analyzed using the example of United States and India. According to one report, the United States's IP is worth $5.8 trillion that is more than the GDP of some of the other countries in the world. Among that IP-intensive industries are around 38% of total GDP of the United States. IP also accounts for 74% of exports in the United States [17]. An overall analysis depicts that the direct and indirect economic impacts of invention are quite effective which is responsible for more than 40% of U.S. economic growth and employment. In order to analyze the role of IPR protection in R&D, which in turn leads to economic growth, an analysis was carried out that revealed that if IPRs would not have been protected in the United States, 60% of pharmaceutical industries and 40% of chemical industries would not have been developed.

Discussing about a developing nation like India, efforts are being made towards development of policies regarding Indian IP system. According to a report published in 2013, it's the advancement of inventions through research and development in the country that has established the Indian IP status in the world. It is believed that the growth in Indian economy is an effect of growing IP impact in the country. Major steps were taken in India regarding IPR in the TRIPS agreement, complying the U.S. and European IPR structure [18]. Since the establishment of patent Law in India, many reforms have been made and according to the survey in 2013, there is 20 times increase in Trademark and Patent filings in comparison towards the previous years. This has increased the interest of foreign companies in India and many public and private R&D centers have been built up.

Recently, in May 2016, government has approved the National IPR Policy that will give the direction to the future of IP in India. By understanding the role of IP in economic growth, government is now taking initiative towards promoting IPR through outreach activities, to strengthening the Legal and Legislative Framework, strengthening the enforcement and adjudicatory mechanisms, and so on.

It has been assessed that developing nation like India needs to improve the IP framework mainly regarding protection of life sciences IP, new enforcement mechanisms, and specialized IP courts. This requires new procedures for local production, procurement, and manufacturing, and so on [19, 20].

8.4.1 Valuation of IPR

As it has been understood by now, how IPR is directly related to economic growth, thereby its importance has increased over the years. Because globalization is increasing at a very rapid rate in all spheres, therefore, it has become increasingly important to get the invention evaluated. The global economic environment has led to the formulation of new models for international and national trading, where valuation of IPR is an integral part. Thus, the importance of the valuation of IPR is because of the following reasons:

- To protect invention from exploitation
- To raise funds

- For sharing of profit (in case of multiple IP owners)
- To save taxation
- For assessing damage claims in a dispute, infringement or breach of contractual rights
- To assess transfer pricing of patents or licensing a trademark

The valuation of IP is done using cost, market, and income [21].
These methods can be used individually or using a combined approach where all the three methods are utilized concurrently.
The various factors that should be kept in mind while valuing IPR are as follows

- IP should be clearly identified
- Unambiguous title should be avoided
- Qualitative and quantitative characteristics of the IP
- IP should be assessed in terms of earnings capacity and profitability
- IP has indulgence in Market share
- Legal rights and restrictions, and risks associated with the IP
- Historical growth and prospects for the future

8.5 Role of Intellectual Property in Technology Transfer

Technology transfer in simple words is the transfer of research/developments in to a product form which can be commercialized. Due to constantly growing level of competitive passion, technological expertise has occurred as some of the chief motivating competencies for incredible corporate and new-age start-ups. The fast developments in technological expertise have redefined the establishments' function and behaviour of industry throughout zones. In this digital age, science has become invaluable thus allowing infringers to easily counterfeit any creation, therefore decreasing the originator's incentive to involve in such routine. IPR supports the inventors to look after their creations and to prohibit others to create, trade, or fabricate for a period of predefined years. IPRs permit innovators to get worthwhile skills on the market, for the reason favorable to them for his or her progressive struggles and reward them for the funds acquired in the course of the study and progress.

An adequate and strong IPR policy aids setting up nations in growth and science transfer, therefore benefiting for novelty and offering revenues in research and progress.

8.5.1 IPR and Licensing

Licenses will also be within the type of price, royalty or profit-sharing groundwork, as a result delivering the right to supply or fabrication of the product for a given interval of time within a distinctive area. Organizations giving inferior technology products involved in licensing as compared to other routes corresponding to FDI. Sometimes as a part of strategy, competitor party can be prevented from obtaining patent by defensive publication. Superior IPR defense can slash licensing price. It additionally discourages novelty available in the market. Thus, IPR defense has a constructive influence in licensing.

8.6 The Philosophy of Intellectual Property Rights

IPR maintains the establishments of the human brain and intellect. The intellectual property rights method has been developed in order to achieve two clashing aims:

- To advertise the e-newsletter of recommendations, innovations and creations, with the intention to make them on hand to others, who can then give an extra boost to them, which may increasingly nurture scientific development or inventive proposal
- To stipulate an economic incentive for public to formulate or to engage in ingenious works, via making certain that the inventor can obtain monetary rewards from his/her efforts

The answer accepted was to present the discoverer or creator a provisional monopoly, in trade for making his/her concept known to humanity.

A patent needs the inventor to reveal his/her innovation, in conversation with a provisional monopoly on its use. On account that of this invention, the originator will likely be capable to receive a profit in case of commercialization of the invention, both via direct utilization or through royalties, further giving a license to make use of the invention. Creation of the similar product by the way of a dissimilar construction method however does not contempt an approach patent.

8.7 IPR as an Instrument for Development

IPR security helps inventors to gain incentive to invent and inspire long term development. Strengthening the IPR management can enhance country's development and results in better growth in open market. Enhanced IPR safety may additionally support in decreasing imitation, which leads to scale back competitors and inspire innovation within the domestic market. There is a lined relationship between IPR and development as IPRs are destined to affect trade leading to higher trade flows for industries as excessive technology or patent sensitive.

8.8 Some Examples of Patent

8.8.1 Self-Healing Power Grid and Method Thereof (U.S. Patent No. 8504214)

Smarter electrical programs are being developed with the aid of more than a few manufacturers involved in engineering electrical techniques on the building or grid stage. Some of these improvements deal with the efficient use of electrical energy, including programs of drawing power to a building from a grid in a way that takes top demand hours into account, establishing more price-mighty vigor programs. These trends have the capability to largely alter vigor use foundation.

The USPTO issued this patent to the overall electric organization of Niskayuna, New York, in August 2013 to safeguard a self-medication electrical grid that may maintain small malfunctions from causing main grid outages. A grid monitoring procedure judges the infectiousness fee of the add-ons making up the grid, which is a measure of the chance of a single element breaking down and creating an excessive load on neighbouring components. If the infectiousness price passes a special threshold, the grid can alter the running capacity of the add-ons to avoid better grid issues from establishing.

8.8.2 Infectious Disease Detection (U.S. Pat. App. No. 20130130227)

In an increasing world, air travel takes core stage because the main mode of transportation for business gurus and vacationers are alike. As the global leader in aviation, the Federal Aviation Administration (FAA) owe to engage internationally for increase in global safety standards and enhance aviation safety and efficiency. Some of the excellent considerations of safe air travel are the ability for handling infectious sickness to journey halfway around the globe within hours, reaching a colossal population where a deadly disease would be disastrous.

This patent software, filed with the USPTO by means of The Boeing manufacturer of Chicago, IL, would shield a procedure competent of deciding upon whether or not a contagious disease is gift in a field of an airport. The approach makes use of quite a lot of sensors that may become aware of contagion stages of a disorder and alert a management system if those phases point out that a severely infectious disorder is gift. This system could be used to observe the presence of tuberculosis, the Ebola virus, influenza or a style of alternative contagions that can be detected via the presence of micro organism, an epidemic, or a different particulate.

8.9 Summary

Awareness, significance, and encouragement toward IPR is exponentially increasing in the research community. This chapter has provided foundations and directions to an IP enthusiast to ponder upon key provisions and various forms of IP such as copyright, patent, trademark, geographical indications, integrated circuit, and so on. The chapter details the procedures of patents and highlights the do's and don'ts. Critical issues such as defense IPR are also briefed. Similarly, the issues related to IP domain encompassing the role of IP in global economy, valuation, and technology transfer, IPR philosophy has been discussed in brief. Examples presented at the end of the chapter provide insights in IPR and validates that role of IPR in overall development ensures integrity and transparency in the growth.

Further Reading

1. www.wipo.int/designs/en/
2. http://www.theglobalipcenter.com/resources/why-is-ip-important/
3. https://www.wto.org/english/docs_e/legal_e/27-trips_04b_e.htm
4. https://www.theglobalipcenter.com/
5. Kaushik Laik, "Role of intellectual property in economic growth", J. Intellectual Property Rights, vol. 10, pp. 465–473, 2005.
6. http://patentscope.wipo.int/search/en/structuredSearch.jsf
7. http://www.adelaide.edu.ac/CIES/0022
8. http://www.thehindu.com/business/Economy/Intellectual-Property-index-India-remains-near-bottom/article17277994.ece
9. https://www.jpo.go.jp/
10. http://www.wipo.int/export/sites/www/sme/en/documents/pdf/ip_panorama_11_learning_pdf
11. http://www.ipindia.nic.in/form-and-fees.htm
12. http://www.wipo.int
13. http://www.ipindia.nic.in/
14. Maskus Keith, Intellectual property rights and foreign direct investment, Policy Discussion paper 0022, University of Adelaide, November 24, 2004.
15. https://www.wto.org/english/tratop_e/trips_e/intel2_e.htm
16. www.wipo.int/wipolex/en/text.jsp?file_id=128089
17. http://sicldr.gov.in/
18. http://www.espacenet.com/access/index.en.htm
19. http://www.ipindia.nic.in/writereaddata/Portal/IPOAct/1_43_1_trade-marks-act.pdf
20. http://www.ipindiaservices.gov.in/publicsearch
21. http://www.uspto.gov/patft/index.html

Index

A

abstracts
  reading, 50
  reports, 46
  scientific papers, 38
acknowledgments, reports, 46
ACM (Association for Computing Machinery), 35
action recognition
  literature survey, **56**
  methods, **57**
Adjoint Sensitivity Analysis Procedure (ASAP), 204
Agrawal, A., 68
Ali, Saad, 67
Allwine, Rochelle, 159
alternative hypothesis (H_a), 19
analysis, 12, 20
  result analysis, 23
  statistical analysis methods, 23
analytical research, 6, 8
Andrade, Chittaranjan, 234
appendices, reports, 47
Apple iPhone, 4
application research, 6
applied research, 6
applied statistics, 160
  bi-variate analysis, 181
  inferential statistics, 174
  modelling, 194–198
  multi-scale modelling, 201
  multivariate analysis (MVA), 182
  multivariate analysis of variance (MANOVA), 183
  parameter estimation, 169
  performance prediction, 199
  principal component analysis, 184
  regression analysis, 161–166
  sensitivity analysis, 204
  state vector machines (SVM), 187
  Statistica software tool, 204
  uncertainty analysis, 190–191
  univariate analysis, 178
  univariate analysis of variance (ANOVA), 182
articles. *See also* primary research article
critically evaluated points, 54

journal papers, evaluating, 32
journals, 14, 34
ASCE (American Society of Civil Engineers), 35
ASME (American Society of Mechanical Engineering), 35
Aubry, Keith B., 159
audience focus, presentation, 231
authorship, research report, 49
automoative applications of digital signal processing, 143

B

background research, 27
  incorporating information, 29
  planning process, 29
  planning questions, 29
  sources, 30
  topic limitations, 27
Badie, Julien, 68
Bairagi, Vinayak K., 159, 206
Band, J., 261
Banji, D., 262
Banji, O.J.F., 262
bar graph, 180
Barkhousen noise, 140
basic research, 6
Bathe, Klaus-Jürgen, 206
Bayesian-based regression, 163
Beasley, J.D., 262
behaviour analysis
  literature survey, **56**
  methods, **57**
Bhanu, B., 67
Bhosle, Anand S., 235
bibliographies, plagiarism and, 251
bibliography tools, 228
Bibme research tool, 229
biological engineering, measurment instruments, **151**
bi-variate analysis, flow, 181
block schematic, scientific papers, 39
Boulay, Bernard, 68
Bremond, Francois, 68
Briggs, Forrest, 159
Brown, John M., 159

C

Cacuci, Dan Gabriel, 206
calibration of instruments, 115
Carey, Andrew B., 159
Carr, Joseph J., 159
causal relationships, explanatory research design, 76
certification marks (CM), 280
Cha, H., 157
Chai, Y., 67
Chang, K., 67
cheimcal engineering, measurement instruments, **157**
Chen, Q., 67
Chen, Y., 67
chi-square test, 20
    inferential statistics, 174
Chiverton, J., 67
Choi, H., 146
Choi, S., 146
Choudhari, K.K., 159
citations
    plagiarism and, 251
    scientific papers, 40–41
Citavi research tool, 229
Citefast research tool, 229
Citelighter research tool, 229
civil engineering, measurment instruments, **154**
Clarivate Analytics, 35
classification, 19
cluster sampling, 95
collective marks (CM), 280
colwitz research tool, 229
comparative scaling, 137
completely randomized experiment design, 83
    random replications design, 83
    two group design, 83
complex relationships, explanatory research design, 76
componental theory of individual creativity, 5
computer engineering
    measurement instruments, **153–155**
    topics, 12
conclusions, 49
    reports, 47
    research design, 73
    research report, 47–48
    scientific papers, 40
conference papers, 34
confounded relationships, 73

contact noise, 140
control groups, 73
    pre-test post-test only design, 81
Controller Area Networks (CAN), 143
controls, 73
convenience sampling, 97
copyrights, 279
correlational studies, 3, 7
Creative Commons (CC) license, 261
Creative Commons Attribute (CC BY), 261
Creative Commons Attribution - Non Commercial - No Derivs (CC BY-NC-ND), 261
creativity, 4
    examples, 4
    individual, componental theory, 5
    workgroups, 6
Crewson, Phil, 206
Cristianini, Nello, 206
cross over experiment design, 88
Currie, J.I., 206
Cutler, R., 67

D

Darwin, Charles, 5
data collection, 11, 17–18, 131
    collection method, 17–18
    digital signal processing, 139–143
    primary data, 131
    secondary data, 132
    sources, 132–137
data representation. *See also* classification
    graphical, 19
    methods, 18–19
datasets, studying, 50
Davis, L., 67
Dawson, Catherine, 98
descriptive research, 3, 6–7
    research design, 76
device input, 23
diagnostic research, 3
    research design, 76
digital filtering, digital signal processing, 142
digital signal processing
    automotive applications, 143
    bird song structure case study, 143
    digital filtering, 142
    electrocardiography, 141
    genomics, 142
    geophysical, 141
    medical instrumentation, 142

noise filtering and, 139
noise sources, 140
physiological sounds in cardiac cycle, 144
seismology, 142
structural health monitoring, civil engineering, 142
direct relationships, explanatory research design, 76
discussion
reading, 50
scientific papers, 38
Doebelin, E.O., 159
D-W test, 20
dynamic characteristics of instruments
accuracy, 104
dynamic error, 104–105
fidelity, 104
lag measurement, 104
response speed, 104
span, 105
sphygmomanometer case study, 105

E

Easy Bib bibliography maker, 228
e-books, evaluation, 32
electrical engineering
measurement instruments, **152**
topics, 12
electrocardiography, digital signal processing, 141
electronic databases, 34
electronics engineering, measurement instruments, **145–149**
encyclopaedias/dictionaries
conference papers, 34
electronic databases, 34
Internet, 34
libraries, 34
EndNote research tool, 229
engineering, measuring instruments, **145–157**
environmental effects on reliability, 127–133
environmental engineering, measurement instruments, **153**
errors, sampling/nonsampling, 91
Evangelio, Rub'en Heras, 67
Expected Performance curve (EPC), 200
experiment design, 79
completely randomized, 83
cross over, 88
factorial, 86
Latin Square, 85
post-test only, 81
pre-test post-test only, 81–83
quazi-experimental design, 87
randomized block, 84
experimental groups, 73
experimental research, 9
experimental research design, 77
principle of local control, 80
principle of randomization, 80
principle of replication, 79
experimental setup, 110
case study, field based setup, 112
field based (outdoor), 111–113
instrument calibration, 115
laboratory based (indoor), 111
natural/quasi experiments, 114
steps, laboratory experiments, 115
experimentation, scientific papers, 39
explanatory research, 9
research design, 75–76
exploratory research, 9
research design, 77
extraneous variables, 73
eXtyles, 229

F

factorial experiment design, 86
fail-safe design, 126
Fern, Xiaoli Z., 159
field-based experimental setup, 113
figures, 20
findings. *See also* results
conclusions, 49
qualitative analysis, 41
quantitative analysis, 41
recording, 51–55
reports, 47
research design and, 73
research report, 47
scientific papers, 40
summarizing, 51–55
time performance analysis, 41
flicker noise, 140
focus groups, literature survey, 36–37
formulation of hypotheses, 12
Forward Sensitivity Analysis Procedure (FSAP), 204
frequency distribution table, 179
F-test, 20
fundamental research, 6
further research, scientific papers, 41
future scope, scientific papers, 41

G

Galton, F., 163
gap in research, proposal, 16
Garate, Carolina, 68
Garnier, C., 206
Gates, Bill, 265
genomics, digital signal processing, 142
geographical indications, 279
geophysical data signal processing, 141
Gerstman, Bud Burt, 206
Ghost Writer (plagiarism), 241
Gilbert, Frederick F., 159
Good Man, Erik D., 67
Google Books, 32
Google Scholar, 32–33
Gopal, Vajjha Hara, 205
Grammerly, 229
graphical representation, 19
graphs, 20
    bar graph, 180
    frequency distribution table, 179
    histogram, 180
    pie chart, 181
Grassie, T., 206
Gupta, Santosh, 98
Gyorgyi, Albert Szent, 2

H

H_0 (null hypothesis), 18
H_a (alternative hypothesis), 18
Hansen, Andrew J., 159
Haque, Mahfuzul, 67
Harani, A., 262
hardware, scientific papers, 39
Harpale, Varsha K., 159, 206
Hati, Kalyan Kumar, 67
Henderson, D., 206
Heng, P.A., 67
Hidalgo, Jorge Arturo, 206
histograms, 180
Houghton, Andrew R., 159
Hufftech, Mark H., 159
humidity, reliability and, 127
Hunter, Matthew, 159
hypotheses, 3
    formulating, 10, 15
    null hypothesis (H_0), 19
    research design and, 72
    testing, 11, 19–20
hypothesis testing
    alternative hypothesis, 19
    chi-square test, 20
    D-W test, 20
    F-test, 20
    limitations, 20
    null hypothesis, 19
    steps, 20
    t-test, 20
    Type I errors, 19
    Type II errors, 19
hypothesis-testing research design, 78

I

identification of keywords, 11, 13
idustrial design, 270
IEEE electronic library, 35
IEEE Explore, 35
impact factor journals, 35
indirect relationships, explanatory research design, 76
industrial property, 265
inertia of large numbers, sampling and, 89
inferential statistics, 174
    chi-square test, 174
    T-test, 174–176
instrument calibration, 115
instrument characteristics
    dynamic, 104–105
    static, 100–105
instrument reliability
    availability and, 119
    characteristics, 119
    cost and, 126
    designing for, 125
    ECG (electrocardiogram) case study, 128
    environment effects, 127–130
    failure density and, 122
    failure rate, 117
    function and failure rate calculation, 122
    maintainability, 123
    managing, 119
    medical product case study, 128
    metrics, 118
    MTBF/MTTF and, 122
    non-repairable systems, 119
    quality and, 118, **118**
    repairable systems, 119–120
    temperature and environmental case study, 127
    Weibull++ and, 123
instrumentation, 99–100
    data collection, 131
    engineering measuring instruments, **145–157**

Index

experimental setup, 110–115
instrument characteristics, 100–105
instrument reliability, 115
medical, digital signal processing, 141
scaling, 137
instrumentation engineering, measurement instruments, **151**
instrumentation schemes
    block diagram, 106
    complex (heat exchanger), 106–108
    simple (oxygen cylinder), 106
integrated circuits, 283
intellectual property (IP), 266
    versus industrial property, 265
    licensing and, 284
    technology transfer and, 286
intellectual property rights (IPRs), 266
    collective marks, 280
    copyrights, 279
    development and, 287
    geographical indications, 283
    in global economy, 284
    industrial design, 282
    integrated circuits, 283
    oppositions, 278
    patents, 267–274
    philosophy of, 287
    significance, 265
    trademarks, 280
    valuation, 285
intelligence
    practical, 4
    synthetic, 4
international patent applications
    Paris Convention, 276
    Patent Cooperation Treaty (PCT), 276
    post-grant opposition, 278
    pre-grant opposition, 278
Internet, 34
interpretation section, 51
interpretations, research design, 74
interview methods, 23
interviews
    data collection and, 139
    semistructured, 37
    structured, 36
    unstructured, 36
introduction
    reports, 46
    scientific papers, 39
iPhone, 4

J

Jayanth, J., 1, 206
JCCC (J-gate Custom Content for Consortia), 35
JCCC@JNDEST, 35
Jodoin, Pierre-Marc, 67
Johnson, Burke, 206
Jonson, Ben, 239
journal citation reports (JCR), 35
journal papers, evaluating, 32
journalism, plagiarism in, 247
journals
    ACM (Association for Computing Machinery), 35
    ASCE (American Society of Civil Engineers), 35
    ASME (American Society of Mechanical Engineering), 35
    IEEE electronic library, 35
    IEEE Explore, 35
    impact factor journals, 35
    JCCC (J-gate Custom Content for Consortia), 35
    JCCC@JNDEST, 35
    Math SciNet, 35
    Nature journal, 35
    Open Access, 14
    reputed journals, 35
    Science Direct, 35
    SciFinder scholar, 35
    SCOPUS, 34
    Springer publishing company, 35

K

Kafai, M., 67
Kale, Geetanjali V., 1
Keith, Maskus, 289
keyword identification, 11
Kim, T., 67
Konrad, Janusz, 67
Kothari, Chakravanti Rajagopalachari, 24, 98, 159
Kumar, Ranjit, 24, 98

L

laboratory-based experimental setup, 111
    steps, 115
Laik, Kaushik, 289
LaTeX, 229
Latin Square experiment design, 85
Levy, Robert, 206

Li, Dawei, 66
Li, S., 66
libraries, 34
Lin, B., 67
linear ordinary least square (OLS) regression, 166
linear scaling, 139
literature review, 11, 14–15
    content development, 54
    organizing, 55
    presentation, 15
    proposal, 16
    sources, 14
    survey steps, 14
    thesis location, 54
    writing, 55
literature survey
    action recognition, **56**
    behaviour analysis, **56**
    encyclopaedias and dictionaries, 34
    findings, recording and summarizing, 51–63
    focus groups, 35–37
    interviews, 35–36
    journal citation reports (JCR), 35
    object classification, **56**
    object detection, **56**
    object tracking, **56**
    online information access method, 31
    parts, 27
    patents, 43–45
    points, 15
    questionnaires, 35, 38
    research design and, 71
    research reports, 46
    resources, 30–37
    scientific papers, 39 reading, 38–41
    snowball method, 31
    systematic manual search, 31
    white papers, reading, 42
Liu, Si, 67
local control, experimental research, 80
Local Interconnect Network (LIN), 143
Lu, Hanqing, 67

M

machine learning, sampling and, 97
Maenaka, K., 159
Majhi, Banshidhar, 67
Mammone, Alessia, 206
Mane, Pradeep B., 103

Math SciNet, 35
mathematical modelling, 195
    model classification, 194
    modelling process, 196
    modelling stages, 194
    solar water heater example, 198
Mccomb, William C., 159
MCUs (microcontrollers), 143
mechanical engineering
    measurement instruments, **152**
    topics, 12
mechanical vibrations/shock, reliability and, 128
medical instrumentation, digital signal processing, 142
Mendeley bibliographic tools, 228
method of maximum likelihood estimate, 171
method of moments, 170
methodology
    proposal, 16
    research, *versus* research methods, 22–23
methods
    reports, 46
    research, 22–23
    research report, 47
Mirmehdi, M., 67
MIT (Massachusetts Institute of Technology), plagiarism and, 240–241
Mizner, Wilson, 238
modelling
    mathematical, 195–196
    performance measurement, 199
monetary requirements, samples, 90
multiple linear regression, 167
multi-scale modeling
    linear/nonlinear analysis, 202
    structures, 201
multistage sampling, 95
multivariate analysis (MVA), 182
multivariate analysis of variance (MANOVA), 183
Muneer, T., 206
Munot, Mousami V., 235
Murshed, Manzur, 67
Murty, D.V.S., 159

N

Nagmode, Manoj J., 25
Nakra, B.C., 159
national patent applications
    complete specification, 270
    provisional, 270

natural/quasi experiments, 114
Nature journal, 35
Neal, Lawrence, 159
Nehru, Jawaharlal, 255
Newton, Isaac, 14
Newton's laws, 7
Nikam, Shobha S., 99
NinjaEssays research tool, 229
noise (digital signal processing)
    atmospheric, 141
    Barkhousen noise, 140
    contact noise, 140
    electrostatic discharge, 141
    environmental, 140
    external sources, 140
    flicker noise, 140
    internal sources, 140
    power lines, 141
    radiation, 141
    shot noise, 140
    sparking, 141
    thermal noise, 140
    white noise, 140
noncomparative scaling, 137
nonparticipant-based observation, 134
nonprobability sampling, 96
nonsampling errors, 91
Noodle Tools, 229
null hypothesis (H_0), 19
    Type I error, 19
    Type II error, 19

O

object classification, literature survey, **56**
object detection, literature survey, **56**
object tracking, literature survey, **56**
objective research, 6
objectives
    proposal, 16
    redefining, 15
    research design and, 71–72
observation, 23
    data collection and, 132–133
O'Malley, Austin, 12
online information
    e-books, 32
    Google Books, 32
    Google scholar, 32
    journal papers, 32–33
    websites, 33
    Wikipedia, 32
Open Access journals, 14

outcomes, proposal, 19
over-specification, design and, 125
Oxford University plagiarism rules, 246

P

Pandit, Mak, 42
papers, shortlisted, 14
parameter estimation, 169
    estimator properties, 170
    method of maximum likelihood estimate, 171
    method of moments, 170
    Null Hypothesis test of Significance (NHTS), 170
    performance parameter, 172
    point estimate, 170
    sampling distribution of the mean (SDM), 171
parameters
    identification, 10, 16
    samples, 89
paraphrasing, plagiarism and, 252–257
participant-based observation, 133
Patent Cooperation Treaty (PCT), 45
patents, 14
    applicants, 276
    application, 43, 268
    application types, 269–273
    benefits of application, 271
    characteristics, 281–286
    claims, 43
    examination, 44
    filing considerations, 269
    filing in India, 269
    filing procedure tips, 269
    infectious disease detection example, 288
    international applications, 45, 275
    litigation, 43
    national applications, 45, 275
    novelty, 266
    patent pending, 44
    prosecution, 43
    regional applications, 45
    regional office jurisdiction, 274
    searches, 43
    sections, 48
    self-healing power grid example, 287
    specification, 43
    utility, 268
Paul, Vogt W., 206
philosophy of intellectual property rights (IPRs), 287
pie chart, 181

plagiarism, 21, 235
   academia, 245
   accidental, 237–242
   of authorship, 242
   avoiding, 251–257
   bibliography and, 255
   citations and, 256
   COPE (Committee of Publishing Ethics), 254
   Creative Commons (CC) license, 261
   cyber plagiarism, 237
   definition, 241
   digital plagiarism, 237
   Elsevier, 248–253
   forms, 246
   free software, 247
   government rules/policies, 254
   of ideas, 243
   IEEE (Institute of Electrical and Electronics Engineers), 252
   iThenticate software, 247
   journalism, 247
   music industry, 244
   paraphrasing and, 258
   patchwork plagiarism, 237
   programming code reuse, 243
   reasons for, 235–237
   references and, 255
   self-plagiarism, 237
   Springer, 248
   stitching sources (potluck plagiarism), 237
   Turnitin software, 248
   Urkund software, 250
   of words, 237
point estimate
   bias, 170
   consistency, 170
   efficient estimator, 171
polymer engineering, measurment instruments, **153**
Poppe, Ronald, 67
population, samples, 89
posters, 232
Post-it Notes, 4
post-test only experiment design, 81
practical intelligence, 4
presentation, 206
   audience focus, 231
   explanation, 231
   posters, 232
   reports, 229–243
   research papers, 228–234
   slides, 230
   theses, 228–230
   tools, 228
pressure variations, design and, 126
pre-test post-test only experiment design, 81–83
primary data, 131
primary research articles, reading, 50
primary sources, background research, 30
principal component analysis (PCA), 184
   computing principal components, 184
Prins, Jack, 206
probability distribution function (PDF), 190
probability sampling, 92
problem statement. *See also* research problem
   background research, 27
   characteristics, 64
   defining, errors, 66
   defining, method used, 61–62, 64–66
   defining clearly, 61
   formulating through discussion, 62–63
   formulation, 56
   limitations, 59
   literature available, 62
   research design and, 71–72
   selection sequence, 60
procedures
   reports, 46
   research report, 47
publications, 14. *See also* reports; research papers; theses
   abstract, 210
   acknowledgements, 212–215
   articles, 209
   author name and affiliation, 209
   authorship, 217
   bibliography tools, 228
   body of research paper, 207–208
   citations/citation index, 222–225
   conclusion, 212
   conflict of interest, 215
   contents, 210
   documentation tools, 228
   experiment results/discussions, 214
   figures, 215
   future scope, 215
   introduction, 210
   journal identification, 216–217
   journal metrics, 221
   journal ranking, 221
   journal selection guidelines, 216
   letters, 209
   literature review, 213
   methodology, 212

Index   299

open access journals, 220
perspectives, 210
presentation tools, 228
rapid communications, 210
references, 216
research paper identification, 220
research report, 48
review papers, 210
review system, 218
short communications, 210
subscription-based journals, 220
tables, 215
title, 209
publishing, 206
Pulitzer, Joseph, 208

Q

Qiqqa research tool, 229
qualitative analysis, 41
qualitative research, 6, 8
    research design and, 74
quantitative analysis, 41
quantitative research, 6, 8
    research design and, 74
quazi-experimental design, 87
questionnaires, 23
    data collection and, 139
    literature survey, 39
    mailed, 136
    sample, 137
quota sampling, 97

R

Raich, Raviv, 159
Randolph, Justus, 234
randomization, experimental research, 80
randomized block experiment design, 84
Ranjetha, P., 206
Raphael, Martin G., 159
ReadCube research tool, 229
reading methodology, scientific papers, 40
redefining research problem and objectives, 11, 15
redundancy, design and, 125
RefDot research tool, 228
references
    plagiarism and, 258
    reports, 47
    scientific papers, 40–41
RefWorks research tool, 229
Registered trademarks (TM), 280

regression analysis
    Bayesian-based, 163
    effective relationships, 163
    regression model, 161–167
regression model
    independent variable, 163
    linear ordinary least square (OLS) regression, 166
    multiple linear regression, 167
    response variable, 163
    simple linear regression analysis, 161–164
    unknown parameters, 163
reliability of instruments, 117
    availability and, 119
    characteristics, 119–131
    failure density and, 122
    failure rate, 117
    function and failure rate calculation, 122
    maintainability, 123
    managing, 119–120
    metrics, 118
    MTBF/MTTF and, 122
    non-repairable systems, 119
    quality and, **118**
    repairable systems, 119–120
    Weibull++ and, 123–124
Ren, X., 66
replication, experimental research, 79
reports. *See also* publications: research report
    abstracts, 47
    acknowledgments, 47
    appendices, 48
    conclusions, 47
    contents, 210–220
    introductions, 47
    methods, 47
    presentation, 228–230
    procedures, 47
    references, 48
    results (findings), 47
    review of literature, 47
    summary, 47
    table of contents, 47
    title, 46
representation of data, 11, 17. *See also* presentation
reputed journals, 35
research design, 14, 17, 69–70
    conclusions, 74
    data, 73
    descriptive research, 77
    diagnostic research, 77
    experiment design, 79–87

experimental research, 77
explanatory research, 75–76
exploratory research, 78
framework, 71
hypothesis-testing, 78
importance, 72
interpretations, 74
literature survey, 72
method, 74
necessity, 71
objectives, 72
parameters, 71–74
problem statement, 71–72
qualitative research, 75
quantitative research, 75
results, 74
sampling, 88–91
scope, 72
terminology, 72
title, 71
research methods
    device input, 23
    interview method, 23
    observations, 23
    questionnaires, 23
    result analysis methods, 23
    statistical analysis methods, 23
    survey methods, 23
research papers. *See also* publications: contents 209–214: presentation 227–230
research problem. *See also* problem statement
    characteristics, 64
    defining clearly, 61
    definition, 57–62
    formulation, 12, 12–13
    redefining, 15
    research statement, 14
    SMART (Specific, Measurable, Achievable, Realistic, and Time-bound), 14
research process, 10
    data collection, 11, 17–23
    data representation, 19
    formulating hypothesis, 11, 15
    hypothesis testing, 11, 19–20
    identification of keywords, 11, 13
    identification of variable/parameters, 14, 14–17
    literature review, 11, 14–15
    patent, 14
    preparing research proposal, 11, 15–16
    publication, 14
    redefining research problem and objectives, 11, 15
    research design, 14, 17
    research problem formulation, 10, 12–13
    research report, 11, 21–25
    results and analysis, 20
    selection of domain, 11–12
    testing of proposed design on collected data, 11, 19–20
    thesis writing, 11
research proposal, 16
    format, 16
    preparation, 10, 15–16
    sections, 16
research report, 11, 21, 46. *See also* reports
    authorship, 49
    components, 47
    conclusions, 47–48
    findings, 47
    format, 20–21
    hypotheses, 47
    methods, 47
    plagiarism, 21
    primary article reading, 49–50
    procedures, 47
    publishing, 49
    purposes, 21
    sections, 21
    stages, 22 types 54
research statement, 14
    research design and, 71–72
    SMART (Specific, Measurable, Achievable, Realistic, and Time-bound), 14
research analytical, 6–7
    application, 6
    applied, 6
    basic, 6
    correlational, 6–7
    descriptive, 3, 6–7
    diagnostic, 3
    existing theories and, 2
    experimental, 9
    explantory, 9
    exploratory, 9
    features of good research, 23
    fundamental, 6
    importance, 6
    motivation, 3
    objective, 2, 6
    phases, 10–22
    qualitative, 6, 8
    quantitative, 6, 8
    types, 5–9
researchers, characteristics, 3–6
resources

Index

evaluating, 52–53
literature survey, 31–39
locating, 51
quality material, 52
selecting, 52
result analysis methods, 23
results, 11, 20. *See also* findings
    conclusions, 49
    qualitative analysis, 41
    quantitative analysis, 41
    reports, 47
    research design and, 73
    scientific papers, 40
    time performance analysis, 41
retrospective approach, research problem selection, 59
review of literature, reports, 47
Roebuck, Alun, 159
Rohatgi, Vijay, 205
Rouse, Magret, 42
Ruggiero, Leonard F., 159
Rushdie, Salman, 255

S

Sa, Pankaj Kumar, 66
safety critical design, 125
Saleh, MD Ehsanes, 205
Saligrama, Venkatesh, 67
sample design, 90
    errors, 91
    preparation, 90
    procedure selection, 91
sampling distribution of the mean (SDM), 171
sampling cluster, 94
    control, 91
    convenience, 97
    cross validation, 97
    errors, 91
    hold-out technique, 98
    inertia of large numbers, 89
    K-fold cross validation, 98
    machine learning approaches, 97
    method of sampling, 90
    monetary requirements, 90
    multistage, 95
    nonprobability, 96
    parameters, 90
    percentile split, 97
    population, 89
    probability, 92
    quota, 97
    sample frame, 90
    simple random sampling, 92
    size, 90
    snowball, 97
    statistical regularity, 89
    stratified, 93–96
    systemic, 93
    unit of sampling, 90
Sandhya, S., 262
Sapre, A., 67
Sarumathi, S., 206
Sathish, K.D., 262
Sawhney, A.K., 159
scaling
    comparative, 137
    linear, 139
    noncomparative, 137
Schrodt, Philip, 206
Science Direct, 35
scientific papers, 38
    abstract, 39
    block schematic, 39
    citations, 40–41
    conclusion, 41
    discussion, 41
    experimentation, 39
    further research, 41
    future scope, 41
    hardware used, 39
    introduction, 39
    literature survey, 39
    objectives, 42
    reading methodology, 40
    references, 40–41
    results, 40
    software used, 39
SciFinder scholar, 35
SCOPUS, 33, 34
secondary data, 132
secondary sources, background research, 30
seismology, 142
selection of domain phase, 10–12
sensitivity analysis, 204
service marks (SM), 281
Shah, Mubarak, 67
Shakespeare, William, 239
Shanthi, N., 206
Shin, S., 67
shortlisted papers, 15
shot noise, 140
Sikora, Thomas, 67
simple linear regression analysis, 161–164
simple random sampling, 92
Singh, Khushwant, 255

slides, 230
SMART (Specific, Measurable, Achievable, Realistic, and Time-bound), 14
snowball method, 31, 97
software, scientific papers, 39
sources
    background research, 29–30
    data collection, 132–139
    evaluating, 52–53
    finding, 51
    literature review, 14
    quality material, 52
    selecting, 52
Springer publishing company, 35
Srinivasan, Sesha A., 235
state vector machines (SVM), 187
static characteristics of instruments
    accuracy, 102
    drift, 103
    hysteresis, 103
    precision, 104
    range (span), 103
    repeatability, 103
    reproducibility, 103
    resolution, 103
    sensitivity, 103
    stability, 104
    tolerance, 104
Statistica software tool, 204
statistical analysis methods, 23, 178
statistical regularity, sampling and, 89
statistics. *See also* applied statistics
Sternberg, Robert, 4
Sternberg, Robert J., 24
stratified sampling, 93–97
structural health monitoring, civil engineering, 142
structured interviewing, 135
studies, correlational, 3
summary, reports, 47
Sun, Q.S., 67
survey methods, 23
synthetic intelligence, 4
systematic approach, research problem selection, 60
systematic manual searches, 31
systematic sampling, 93

T

table of contents, reports, 47
tables, 20
Taha, Ahmed, 68
Tang, Nick C., 68
targeted approach, research problem selection, 60
Taylor, Andrew, 159
technology transfer, 286
temperature, reliability and, 127
terminology, 2
    confounded relationships, 73
    control, 73
    control group, 73
    experimental group, 73
    extraneous variable, 73
    hypothesis, 73
    variables, 73
testing of proposed design on collected data, 11, 19–20
thermal noise, 140
theses. *See also* publications: contents 209–214: location in literature review 54; presentation 227–230; writing 10
Thiel, David V., 24
Thomson Reuters, 35. *See also* Clarivate Analytics
time performance analysis, 41
title, reports, 46
topic of research
    limitation, 27
    proposal, 16
trademarks, 280
    certification marks (CM), 282
    collective marks (CM), 282
    Registered (TM), 280
    service marks (SM), 281
T-test, 20
    inferential statistics, 174
Turchi, Marco, 206
Type I error, null hypothesis, 19
Type II error, null hypothesis, 19
Typeset, 229

U

uncertainty analysis
    calculation example, 193
    linear and non-linear systems, 191
    measurement and, 191
    probability distribution function (PDF), 190
    probability theory, 190
    UNICORN software, 194
UNICORN software, 194
univariate analysis, 178
    bar graph, 180
    central tendency, 179

Index

dispersion, 179
  frequency distribution table, 179
  histogram, 180
  pie chart, 181
univariate analysis of variance (ANOVA), 182
unstructured interviewing, 135

V

variables, 73
  identification, 10, 16–17
variations in pressure, design and, 128
Vega, Robyn, 159
Velcro, 4
Vidhya, S., 206
Vinod, K.R., 262
Vishwakarma, S., 67
visual formats, 20
von Braun, Wernher, 2

W

Warhade, Krishna, 206
web of science (WOS), 35
website evaluation, 33
Weibull++, 123–124
Wesley, John, 26
white noise, 140
white papers
  authoring, 41
  contents, 47
  definitions, 42
Wikipedia, 32
Woody, Clifford, 24
workgroups, creativity, 6
World Intellectual Property Organization (WPO), 45
Wright brothers, 13

X

Xia, D.S., 67
Xiang, S., 67
Xie, X., 67
Xu, Changsheng, 67
Xu, Lihong, 67
Xu, Y., 67

Y

Yuan, X., 67

Z

Zaidenberg, Sofia, 68
Zayed, Hala, 68
Zhang, L., 67
Zhang, Tianzhu, 67
Zotero research tool, 228

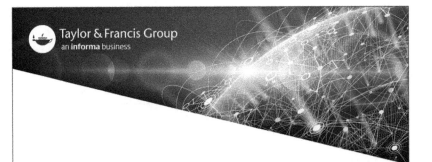

For Product Safety Concerns and Information please contact our EU representative GPSR@taylorandfrancis.com
Taylor & Francis Verlag GmbH, Kaufingerstraße 24, 80331 München, Germany

www.ingramcontent.com/pod-product-compliance
Ingram Content Group UK Ltd.
Pitfield, Milton Keynes, MK11 3LW, UK
UKHW030828080625
459435UK00005B/55